AUTONOMOUS VEHICLES

Your ultimate guide to the
Past, Present and Future
of Autonomous Vehicles

Buckle up and Enjoy the ride !

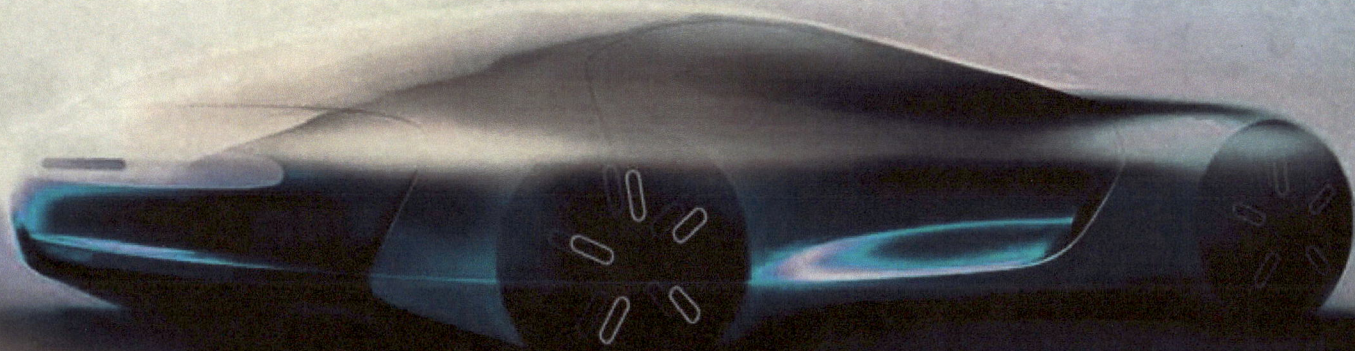

C D LEONARD

CONTENTS

Environmental Perspective

Political Perspective

Socio-Cultural Perspective

LEGAL DISCLAIMER

The information contained in this book and its contents Is not designed to replace or take the place any form of medical or professional advice; and is not meant to replace the need for independent medical, financial, legal or other professional advice or services, as may be required. The content and information in this book has been provided for educational and entertainment purposes only.

The content and information contained in this book has been compiled from sources deemed to be reliable, and it is accurate to the best of the Author's knowledge, information and belief. However, the Author cannot guarantee its accuracy and validity and cannot be held liable for any errors and/or omissions. Further, changes are periodically made to this book as and when needed. Where appropriate and/or necessary, you must consult a professional (including, but not limited to your doctor, attorney, financial advisor or such other professional advisor) before using any of the suggested remedies, techniques, or information in this book.

Upon using the content and information contained in this book, you agree to hold harmless the Author from and against any damages, costs and expenses, including any legal fees potentially resulting from the application of any of the information provided by this book. This disclaimer applies to any loss, damages or injury caused by the use and application, whether directly or indirectly, of any advice or information presented, whether for breach of contract, tort, negligence, personal injury, criminal intent, or under any other cause of action.

You agree to accept all risks of using the information presented inside this book.

You agree that by continuing to read this book, where appropriate and/or necessary, you shall consult a professional (including, but not limited to your doctor, attorney, financial advisor or such other professional advisor) before using any of the suggested remedies, techniques, or information in this book.

INTRODUCTION

CHAPTER

01

UNDERSTANDING AUTONOMOUS

Autonomous driving has been all the rave of late, and some would say years. These days, every single car company seems to have integrated some sort of 'autonomous' feature in their newer models, whilst new car companies seem to have autonomous driving as part of the base of their product offering. However, what is autonomous driving really? And to follow, what separates it from normal driving, hence all the buzz?

The answer to this may be found within the motivations for autonomous driving. The story goes that Karl Benz, the creator of the very first automobile (1885), when taking it on its very first test drive, crashed into a wall. Yikes! And this brought to fore the argument for the least reliable part of the car: the driver. Chris Urmson, former Chief Technology Officer at X, Google's self-driving car team (took over as project lead in 2013) as well as the main engineer that built the code of Google's autonomous software, thinks that the most unreliable part of the car is the driver (see as mentioned, Karl Benz). During his 2015 TED Talks speech, he mentions that humans have tried to correct this part of the car for the past 130 years, adding features such as seatbelts, strength reinforcements as well as airbags. More recently, he suggests that we have also made the car 'smarter' in order to fix the same problem, the human.

The human driver is the most unreliable part of the driving function

Now the CEO of Aurora Innovation, a company that creates autonomous driving software (and recently partnered with Volkswagen and Hyundai in a bid to get autonomous driving commercial as quickly as possible), Urmson continues to explain the difference between a 'patch job' - referring to current efforts to add driver assistance systems to the car - and authentic autonomous driving car development. But for autonomous driving to be truly appreciated, he highlights the scale of the problem, which is that globally, 1.2 million people die in road accidents every year, with 33,000 of those in the United States alone. Back to Urmson a little later on.

AUTONOMOUS DRIVING: DEFINED

What exactly, then, is autonomous driving? Is a car autonomous because due to its ability to drive itself to any destination, with no one inside it? Or is one that takes over in certain situations, but essentially lets the driver engage with the car outside of said situations? Perhaps it involves the vehicle driving itself, without human intervention, as long as certain constraints are met, such as good weather?

SAE J3016 GUIDELINES

The Society of Automotive Engineers ('SAE'), a US based professional association that develops standards targeted towards engineering professionals (with focus placed on transport industries such as automotive, aerospace and commercial vehicles), developed a guide to the levels of automation in vehicles of the future.[1] As part of the Taxonomy and Definitions for Terms Related to On-Road Motor Vehicle Automated Driving Systems standard (SAE International's J3016), autonomous vehicles occupy six levels:[2]

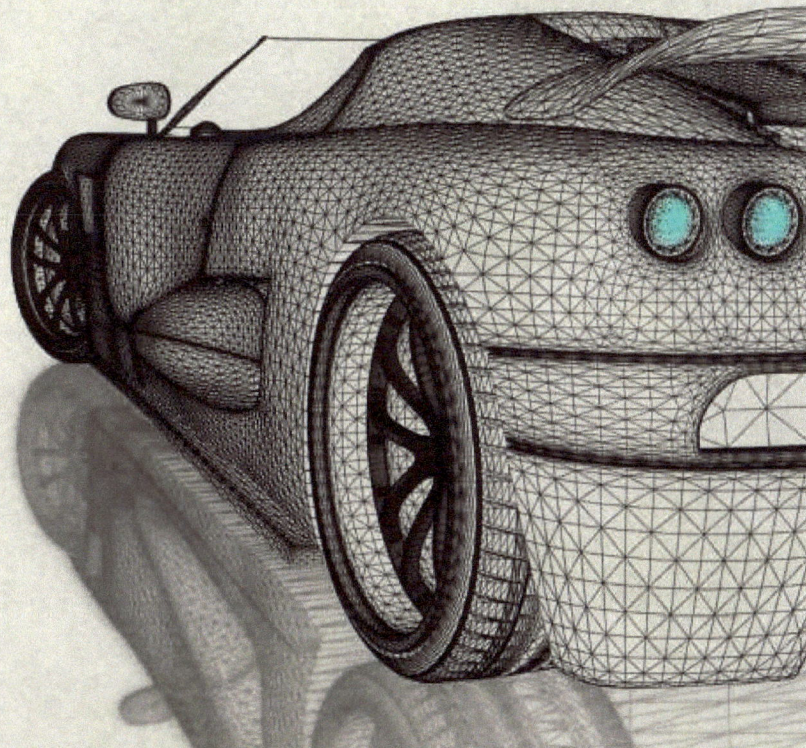

1   Burgess, M. (2017, April 21). When does a car become truly autonomous? Levels of self-driving technology explained. Retrieved from http://www.wired.co.uk/article/autonomous-car-levels-sae-ranking

2   Blain, L. (2017, June 8). Self-driving vehicles: What are the six levels of autonomy? Retrieved from https://newatlas.com/sae-autonomous-levels-definition-self-driving/49947/

What can be more palpably absurd than the prospect held out of locomotives traveling twice as fast as stagecoaches?

The Quarterly Review
March, 1825

SIX LEVELS OF AUTONOMOUS VEHICLES

LEVEL 0

Level 0: No automation - this level of vehicle relies on the human being for all tasks performed. Aspects such as acceleration, braking and steering are in the control of the driver, and the car does not have any system-led interventions. Essentially, you drive it;

LEVEL 1

Level 1: Driver assistance - this car allows for system intervention within certain and specific driver settings and situations, allowing the car to take control of either the steering wheel or the pedals. Further, the system is not in control of both the steering and acceleration/braking at any one time. This level includes features such as park assist and adaptive cruise control, with the key defining characteristic is the driver will most likely be required to have their hands on the wheel most of the time;

LEVEL 2

Level 2: Partial automation - this level features cars that allow the system to take over both the steering wheel and the pedals. However, for the system to trigger such a scenario, a certain condition must be met, such as hands on the wheel within certain times. Also, for such levels, the driver is still required to maintain control of the vehicle, similar to Tesla's Autopilot where the vehicle can conduct some parts of driving whilst the human monitors the environment and performs most of the driving tasks;

LEVEL 3

Level 3: Conditional automation - these vehicles are able to monitor the environment, and crucially, makes some decisions such as when to overtake. However, the driver is expected to be in a position to take control when the system makes a request. The system, in its response to monitoring the environment, can fully take over driving responsibilities. The duality of responsibilities between the developer of the vehicle and the driver makes it easy to see why automakers will be eager to skip this step;

LEVEL 4

Level 4: High automation - this level of vehicle can be driven by a human being, but the human is not required as part of the driving task.[1] The automated system is capable of performing the driving task as well as monitoring the environment, without needing the human to take back control. Under the right circumstances, the car can drive itself full time, can prompt for human assistance, and park itself should the assistance not be provided. Google/Waymo's test cars fall within this category, and can be considered autonomous; and

LEVEL 5

Level 5: Full automation - involves the removal of the human from the driving equation. This level of vehicle would not need any pedals, steering wheels or any other form of human intervention, thus categorized as fully autonomous.

1   Godsmark, P. (2017, October 4). The definitive guide to the levels of automation for driverless cars. Retrieved from https://driverless.wonderhowto.com/news/definitive-guide-levels-automation-for-driverless-cars-0176009/

LEVEL 0

There are no autonomous features.

LEVEL 1

These cars can handle one task at a time, like automatic braking.

LEVEL 2

These cars would have at least two automated functions.

LEVEL 3

These cars handle "dynamic driving tasks" but might still need intervention.

LEVEL 4

These cars are officially driverless in certain environments.

LEVEL 5

These cars can operate entirely on their own without any driver presence.

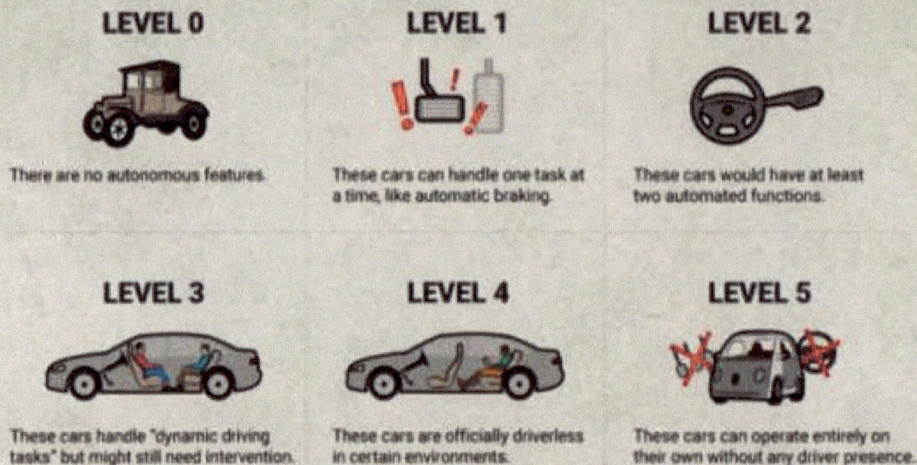

Source: Planetizen [1]

"The automated system is capable of performing all driving tasks, under all conditions that a human driver could perform them."

Being able to monitor the level of automation by vehicle manufacturers, software engineers, or a mix of the two, will be key to understanding how far we truly are from a driverless future. It would also be interesting to be able to understand where Elon Musk's Tesla is truly pioneering autonomous technology, or are Google/Waymo developing the ultimate level 5 system? Perhaps Urmson (the former project lead at Google/Waymo) has some interesting insights into this.

When making the same TED Talks speech, Urmson noted that making incremental additions to driver assisted driving with the hope that this would someday materialise into autonomous driving was like someone that worked really hard at jumping with the hope that they would one day learn to fly.[2] He explains a test that he run alongside other Googlers (Google employees) that were not part of the project team. In it, they trained around 100 Googlers on how to use the vehicle because one of the conditions was that the system still required attention (Level 3, because we live in the real world and accidents do happen). After receiving glowing reviews from the Googlers, Urmson's team decided to playback some of the footage from cameras they had installed in the car, and the team observed that a recurring theme, highlighted most by one example. A gentleman, having noted that his battery was dying, reached out to the back seat of the car for his backpack that had his laptop, which he got out and placed on the driver's seat, before reach-

1   Litman, T. (2017, October 23). The many problems with autonomous vehicles. Retrieved from https://www.planetizen.com/blogs/95445-many-problems-autonomous-vehicles

2   Urmson, C. [TED]. (2015, June 26). Chris Urmson: How a driverless car sees the road [Youtube]. Retrieved from https://www.youtube.com/watch?v=tiwVMrTLUWg&t=56s

ing back again for a charging cable. All the while, the car was hurtling down the freeway at 65 miles per hour.

"The better the technology gets, the less reliable the driver is going to get"

This quote perfectly explains the fundamental problem with incremental improvements to autonomous driving; improvements in the technology may not increase at the pace that the unreliable nature of the driver might increase, and thus producing minimal gains if any. The graph below sums up the differences between a classic car, a driver assisted car, and a self-driving system car.

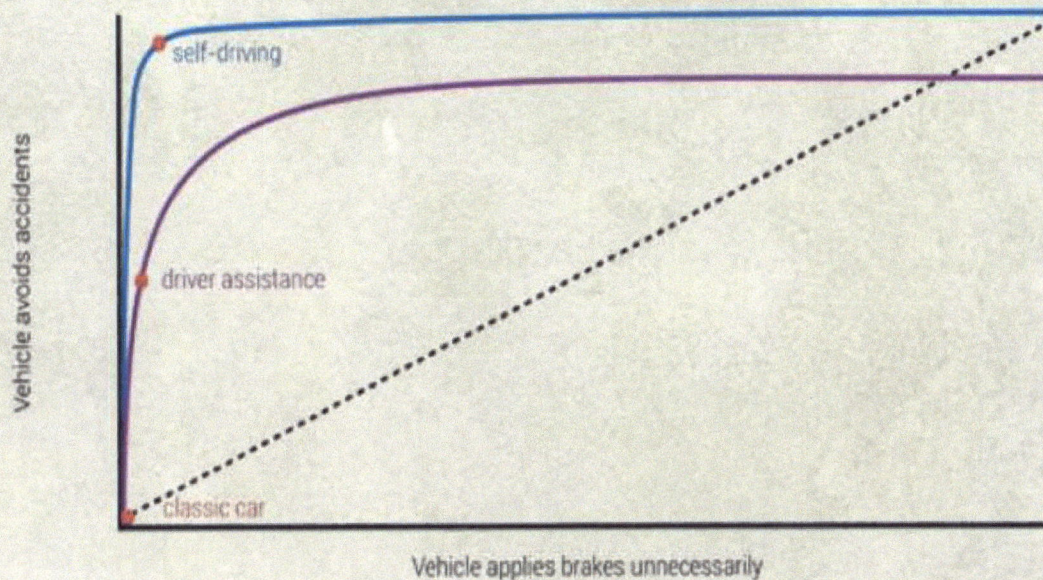

A classic car is one that will have complete human intervention and would thus apply brakes unnecessarily in accordance to the performance of the human being. The driver assisted car is likely to improve the number of accidents stopped without human intervention, with the self-driving car making those accidents even more unlikely. However, the sense of security that the driver may have in the system would mean that it would have to be very good in itself, meaning incremental improvements may not be the strategy to get us there.

That the automobile has practically reached the limit of its development is suggested by the fact that during the past year no improvements of a radical nature have been introduced.

Scientific American
January 2, 1909

AUTONOMOUS DRIVING: NOT HERE YET, BUT HOW FAR?

This is a difficult thing to ascertain. There are several issues to contend with in order to get autonomous over the line and into our lives. For car manufacturers (as well as the software engineers behind them), government approvals across many territories seems to be a big hurdle to get over, as well as the fact that in the US at least, car retention figures seem to be rising.[1] Not only does this mean that people are keeping their cars for longer and longer, but trust in autonomous vehicles within the mass market is also another hurdle to be discussed.

This book will discuss the state of autonomous vehicles today, but not only from a technological perspective, but also from a political, economic, social, legal and environmental perspective ('PESTLE'). These sections will discuss the various developments within the title, as well as anything we might need to look out for in the future.

1   Gelbart, J. (2017, January 4). You may not live long enough to ride a driverless car. Retrieved from http://www.newsweek.com/you-may-not-live-long-enough-ride-driverless-car-575305

TECHNOLOGICAL PERSPECTIVE

CHAPTER

02

TECHNOLOGICAL PERSPECTIVE

"Fully autonomous. There will not be a steering wheel. Twenty years? It will be like having a horse. People have horses, which is cool. There will be people who have non-autonomous cars, like people have horses. It would just be unusual to use that as a mode of transport... I think almost all cars produced will be autonomous in ten years, almost all. It will be rare to find one that is not, in ten years. That's going to be a huge transformation."

~ Elon Musk - CEO, Tesla

Autonomous vehicles are here to stay, if the leaders within the world of driver technologies are to be believed. These technologies may not be as far off as one may think, especially when considering that many of these technologies exist within one form or the other. Think about the brain of the autonomous vehicle with specific thought to the type of data that it needs to receive for it to synthesize to reflect the decisions it will continuously make. Such data will be collected with the use of technologies that already exist today, some of them including the lidars and radars, driver monitoring, dynamic cruise controls and pedestrian detectors, amongst a host of others.

It is only fitting that Elon Musk would think that driverless technologies are bound to dominate the near future, what with all the touted benefits that the technology promises. However, how did we get here? The vehicle, much like the mobile phone, has seen its adoption and influence on large populations and societies grow, albeit perhaps at a slower rate of adoption (scale differences may be explained by differences in price point: cars are significantly more expensive).

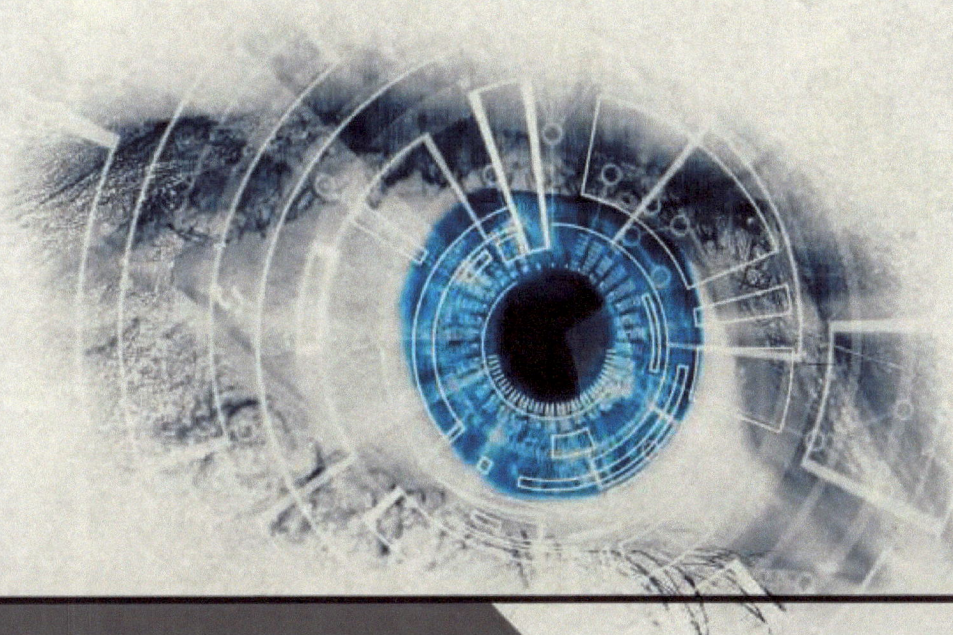

With over fifteen types of foreign cars already on sale here, the Japanese auto industry isn't likely to carve out a big share of the market for itself.

Businessweek
August 2, 1968

THE EVOLUTION OF THE VEHICLE AS WE KNOW IT TODAY

As earlier discussed, Karl Benz is credited with the creation of the first gas-powered vehicle that was later commercialized by Henry Ford. However, in the grand scheme of the evolution of the car from whence it began as a steam-powered vehicle to the form that it currently takes can be attributed to Benz as his creation of the gas-powered automobile sparked the beginning of its evolution. Ford's invention of the assembly line (1906) proved to be the difference however as the invention allowed his company to produce cars at an unprecedented scale, a move that made cars more available (due to lower cost of purchase) to the public due to economies of scale. In addition to this, the implementation of mass production of vehicles meant that the company (Ford) was also able to continually add new features to its vehicles, with some of the first additions being items that include windshields, speedometers, seatbelts and rear-view windows.[1] And yes, you may ask when turn signals were added to vehicles, and the answer would be 1939: they were added to the Buick brand of vehicles, and at the time of their addition, features such as electric windows and air conditioning had already been installed in some vehicles by then.

Beyond their function, cars began to acquire luxury features in the 1950s, a period of time in which features that are now considered mainstream were just being introduced. Consider that some of the earliest installations of power steering technology in vehicles was done around 1951, with other features such as cruise control (1957), three-point seatbelts (1959) and heated seats (1966) were also introduced within this time period. Innovation seems to be a key component within the automotive industry, especially when one considers level of competition. Innovation, a process that allows car makers to introduce features to their vehicles, motivates car manufacturers to beat each other to the introduction of new features to their vehicles. This is so because these features are most often used as selling points for the vehicles themselves, and thus, it only makes sense that car manufacturers would want to include features that are either going to: a) make their vehicle more attractive to similar vehicles in the market from a features point of view; or b) standardise their vehicle when compared to its competitors within the industry.

1   edriving (2018) Evolution of the automobile Retrieved from https://www.idrivesafely.com/defensive-driving/trending/evolution-automobile

THE AUTOMOTIVE INDUSTRY MAY BE IN THE MIDDLE OF A 'REVOLUTION'

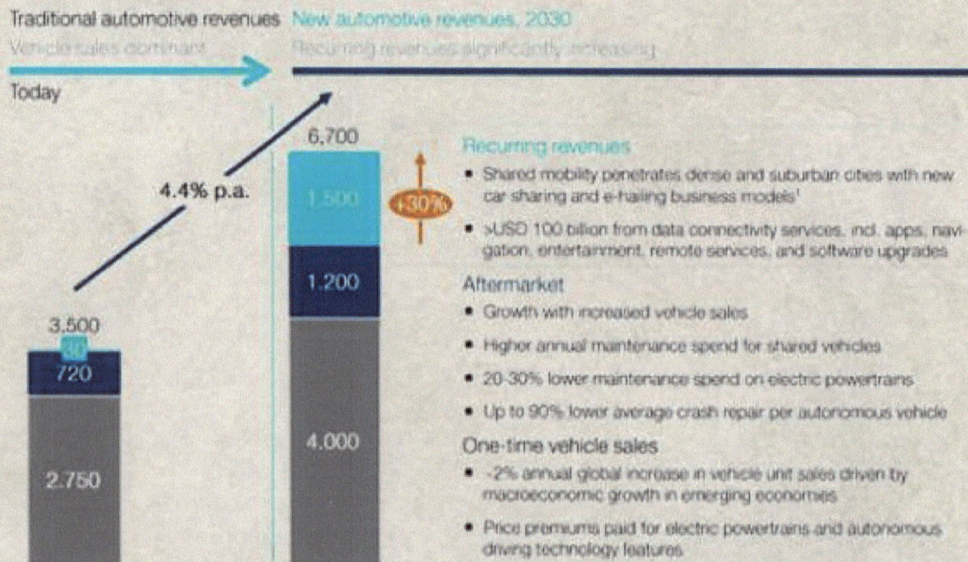

The automotive revenue pool will grow and diversify with new services potentially becoming a ~USD 1.5 trillion market in 2030

USD billions

HIGH DISRUPTION SCENARIO

Traditional automotive revenues   New automotive revenues, 2030

Vehicle sales dominant   Recurring revenues significantly increasing

Today

4.4% p.a.

6,700

1,500

+30%

Recurring revenues
- Shared mobility penetrates dense and suburban cities with new car sharing and e-hailing business models[1]
- >USD 100 billion from data connectivity services, incl. apps, navigation, entertainment, remote services, and software upgrades

1,200

Aftermarket
- Growth with increased vehicle sales
- Higher annual maintenance spend for shared vehicles
- 20-30% lower maintenance spend on electric powertrains
- Up to 90% lower average crash repair per autonomous vehicle

3,500

720

4,000

One-time vehicle sales
- ~2% annual global increase in vehicle unit sales driven by macroeconomic growth in emerging economies
- Price premiums paid for electric powertrains and autonomous driving technology features

2,750

Source: McKinsey

The graphic above, as part of insights gained by McKinsey in their examination of the trends likely to be experienced within the automotive industry, estimate that revenues within the industry are likely to change. The graphic suggests that, although vehicle sales are set to grow at an average of around 2% per annum ('p.a')., total earnings from the entirety of the automotive industry is set to grow at 4.4% p.a. Growth in vehicle sales is explained primarily by demand driven by macroeconomic growth in emerging economies. It is also interesting to note that one-time vehicle sales value is expected to increase from a US $ 2.75 trillion industry globally to US $ 4 trillion partly due to the premium charged on electric power trains (the motors that drive the vehicle) and autonomous driving technology features.[1]

The same report suggests that the aftermarket for automotive revenues is set to increase. However, the real story here is centred on recurring revenues that is

1   McKinsey & Company (2018, January) Automotive revolution – perspective towards 2030. Retrieved from https://www.mckinsey.com/~/media/mckinsey/industries/high%20tech/our%20insights/disruptive%20trends%20that%20will%20transform%20the%20auto%20industry/auto%202030%20report%20jan%202016.ashx

expected to jump to a US $ 1.7 trillion industry from its current US $ 30 billion size. Moreover, whereas recurring revenues currently form approximately 1% of revenues generated by the global automotive industry, these specific revenues are expected to contribute to 22% of total global automotive industry earnings. These are likely going to be a cause of increased penetration of shared mobility services to wider populations including dense and suburban cities. Penetration of these markets are likely to involve the introduction of new car sharing and e-hailing business models. Further, McKinsey also put forward that a data connectivity services sector worth in excess of US $ 200 billion will also be considered part of recurring revenues, and this include applications, entertainment, software upgrades, navigation and remote services.

LET'S TALK ACTUAL TECHNOLOGIES

At its base, driverless technology is a blend of several different technologies grouped together according to function. Therefore, technology with respect to autonomous driving refers to the actual sensors that allow the car to interact with its environment, as well as the other subtleties such as their reliance on rule-based decision engines. Further, we shall also touch on the deployment of large-scale deep learning systems that are crucial to the implementation of large-scale autonomous technologies.

ADVANCED SENSORS FOR MAPPING, LOCALIZATION AND OBSTACLE AVOIDANCE

The first part of the autonomous driving process involves sensing, a key component to the technology. Sensing involves the gathering of data about the environment and thereafter relaying said data for processing. In order to gather data, autonomous vehicles (as currently constituted) utilize several types of sensors. Those technologies that allow localization of the vehicle to centimetre accuracy. LIDAR (Light Detection And Ranging), a laser-based radar, is a remote sensing method that uses light in the form of a pulsed laser to measure variable distances (ranges) to the Earth.[1] These light pulses can be used to generate precise, three-dimensional information about the shape of the Earth, as well as its surface characteristics. LIDAR allows scientists and mapping professionals (critical to the development of autonomous driving technologies) to examine natural and manmade environments with flexibility, accuracy and precision. Though LIDAR is currently expensive technology, several start-ups within the field are working on low-cost LIDAR.

New technologies are also making it easier for autonomous cars to travel in the fog and other obstructed vision scenarios. Previously time of flight systems commonly used in autonomous car prototypes were significantly hindered by poor conditions such as rain and fog. However, researchers at MIT have developed a new system which offers vision at a much higher degree than the human eye.[2]

1   National Ocean Service (2017, October 10). What is LIDAR? Retrieved from https://oceanservice.noaa.gov/facts/lidar.html

2   Krishna, S. (2018, March 22). Autonomous cars may soon navigate better in fog. Retrieved from https://www.engadget.com/2018/03/22/self-driving-technology-fog-mit/

Source: Swapna Krishna

The following image shows how the system developed by MIT is able to better visualise images than the human eye. While the naked eye can typically see through about 30 centimeters of very dense fog, the system was able to see through almost twice that amount. While 60 centimeters is still limited, in practical terms it would be very rare for the density of fog used in the experiment to occur in the natural environment.[1]

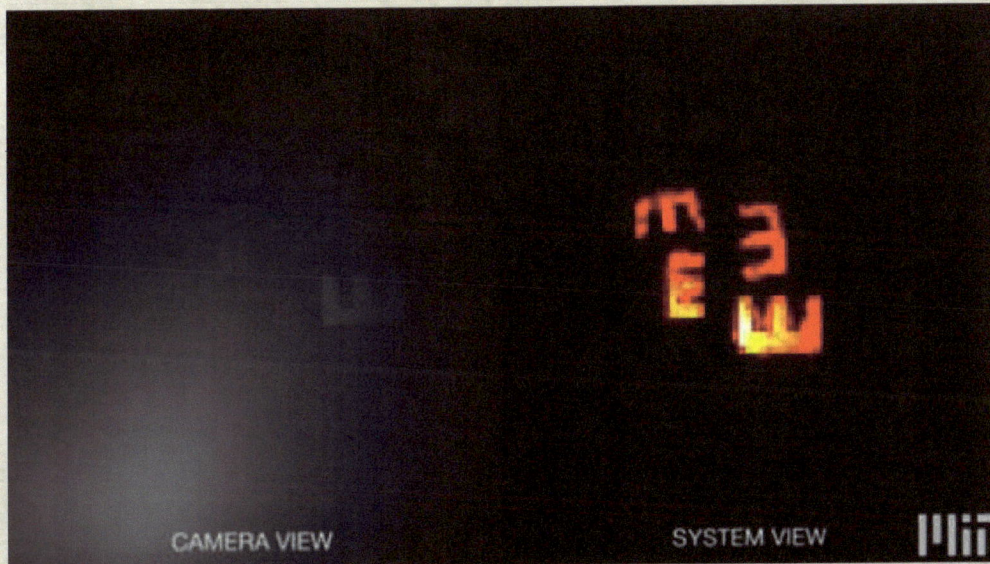

CAMERA VIEW   SYSTEM VIEW

Source: MIT News

1   Hardesty, L. (2018, March 20). Depth-sensing imaging system can peer through fog. Retrieved from http://news.mit.edu/2018/depth-sensing-imaging-system-can-peer-through-fog-0321

OTHER SENSING TECHNOLOGIES

Other sensing technologies include the Global Positioning System (GPS) and the Inertial Measurement Unit (IMU). Experts within the field use a combination of these three technologies in order to localize the vehicle. In addition to these, there is an additional radar sensor that is used for obstacle avoidance and is a failsafe against failure of the three other sensors. In essence, should all the other sensors fail to pick up an object in front of your autonomous vehicle, this sensor kicks into play due to its ability to detect objects that are 5-10 metres away. The sensor is linked directly to the control system of the vehicle and can detect objects in front of it and drive the car away autonomously from the vehicle.[1]

SOPHISTICATED MACHINE LEARNING PIPELINES FOR PERCEPTION

Beyond sensing objects around it, collecting and then relaying the data to aid in decision making, the vehicle is also able to use said data and recognize the objects captured within the data. Therefore, the technology as a whole has a component of deep learning technology commonly used to use camera data to recognize objects around the vehicle. Whilst the sensors on the vehicle may be used to localize the vehicle and figure out how to navigate, object recognition will allow the vehicle to recognize the object data collected by the sensors. The final application for machine learning involves object tracking, a feature that allows the vehicle to track objects (including vehicles) around it. All of this will be made possible through the use of deep learning-based object-tracking mechanisms.

AUTONOMOUS VEHICLES WILL FEATURE RULE-BASED DECISION ENGINES

Driverless vehicles, as part of level 4 & 5 autonomy, will be responsible for taking the human being out of the driving experience either entirely, or by limiting the amount of interaction that the human will have with the vehicle from a driver perspective. In view of this, it is only natural that we expect that the vehicle will be able to make decisions in our place. These vehicles should be able to develop decision pipelines that will guide the sequential decisions that the vehicle will make. Such decisions may include path planning, involving critical questions such as how the car will move from point A to point B, as well as the path that the vehicle will take doing so. In addition, the system must also figure out how the human will issue instructions to the vehicle in order to follow a path desired by the user. Developers of these solutions are likely to use algorithms centred around route planning: The A* algorithm, though often considered impractical, might be a good place to start in figuring out rule-based algorithms.

1. Lorica, B (2016, October 6) The technology behind self-driving vehicles Retrieved from https://www.oreilly.com/ideas/the-technology-behind-self-driving-vehicles

Once the system behind driverless vehicles figures out path planning (as part of their rule-based decision engines, prediction is the next natural step, especially considering that once the vehicle tracks an object, it will be required to use a prediction algorithm based on the results from the tracking. Through the use of the prediction algorithm, the vehicle is able to measure the likelihood of crashing into or avoiding nearby objects. Through the use of prediction algorithms, developers and manufacturers are able to derive the object or obstacle-avoidance decisions required to ensure that the vehicle is able to figure out how to drive away from obstacles or moving objects such that it does not get into an accident.

WHAT DOES THIS ALL MEAN FOR THE FUTURE?

It is increasingly difficult to predict trends, especially considering that technology, a key component of autonomous technologies, are in continuous flux, thus changing continually. However, though we may not be able to fully predict the impact that autonomous vehicles may have on our reality today, we can analyse trends within broader markets and use this for some commentary.

RIDE SHARING WILL INCREASE IN INFLUENCE

According to Business Insider, the continued implementation of autonomous vehicles through to 2025 will likely mean that shared driving will take root globally. The company, whilst collecting data through questionnaires, found that 67% of respondents predicted that the sharing economy will grow significantly, estimating that more people globally will take rides using a sharing service rather than using a privately-owned car. Whilst this may be a bold position to take (it would seem that respondents are bullish on the prospects of autonomous driving), ride sharing is permeating through global societies, and it would not be unreasonable to suggest that it is likely to grow. Add to this the estimate that 10% of all cars in the US are likely to be driverless by 2030, and it would seem that ride sharing, through the use of autonomous technologies, will grow in influence.[1]

COMPETITION WILL BLUR THE LINE BETWEEN THE VARIOUS SEGMENTS IN THE AUTO INDUSTRY

Incumbent players within the sector can be expected to adapt to the changes within the industry. Autonomous technology will be considered a requirement targeted at markets that demand it. Manufacturers will be forced, as we are currently seeing within the market, to collaborate with other companies that may have proficiencies within these technologies or incorporate them in-house. This is especially true of those companies that would like to take advantage of mobility as a service (probably most if not all of them).

Such companies will have to consider competing on multiple fronts, some of which include mobility providers (Uber and Lyft for example), specialty manufacturers (Tesla perhaps) and tech giants (Google and Apple), moves that will eventually complicate the competitive landscape. Companies competing in this space

1   Thompson, C. (2015, November 12). 21 technology tipping points we will reach by 2030. Retrieved from http://www.businessinsider.com/21-technology-tipping-points-we-will-reach-by-2030-2015-11?IR=T#10-of-global-gross-domestic-product-will-be-stored-using-blockchain-technology-2027-21

will not only have to consider the build quality of the product (traditional vehicle characteristics) but also their technological offering, competencies that traditional manufacturers may not presently have. However, companies such as Mercedes, Audi, BMW etc have been relatively successful at developing their own versions of driver assisted technologies, which when paired with fully autonomous technologies, may present them with opportunities for mobility sharing service provision. Point is, the traditional consumer will shift their preferences around the overall product, and this will include technology.[2]

2   Tshiesner, A (2016, March 11). Trends that will transform the auto industry until 2030. Retrieved from https://www.2025ad. com/latest/automotive-revolution/

CONCLUSION

It is difficult to scratch the surface when trying to convey the technology behind autonomous vehicles, let alone explain it in any detail, whilst also keeping it high-level to cover a broader understanding of the concepts. This is how we see it: it is likely that the human being is the most likely to be at fault during an accident, making us very unreliable drivers. Autonomous vehicles, making use of tech such as altimeters, gyroscopes and tachymeters, allow the vehicle to determine the very precise position of the car, and thus due to the high-quality data they produce, allow the car to operate efficiently. Through the use of such technologies, cars will be demonstrably safer as a mode of travel, and this has real consequence on societal impact.

Consider also that technology, by its very nature, is underpinned by change and advancement. It may thus be imprudent to assume that the changing nature of technology is necessarily a bad thing, especially considering human nature generally does not accept change. In this context, the technologies available to humans necessitate most of the changes in particular industries, such as the rise in smart manufacturing. Indeed, change is expected in a multitude of industries as a result of changes in technology, and one may argue that the vehicle is just but one of the many industries affected (whether positively or negatively) by the changes in technology. This does not necessarily mean that change is a bad thing; in fact, as pertains the impact that technology will have on the overall automotive industry, cars are expected to become much safer. Aside from the fact that this means that human lives are likely to be saved, changes in the technology in vehicles is but a part of the overall changing dynamics in the industry. Though inevitable, policy makers are advised to work closely with the developers of autonomous technology to try and mitigate its adverse effects on society.

Autonomous vehicles may change in the nature of their expected benefits, but based on current evidence, they will be able provide humans with free time now that they are likely to be taken out of the driving equation. All this is likely to improve productivity and reduce traffic jams, amongst other potential benefits. Whatever the case, it will be interesting to note the changes that the industry will go through.

ECONOMIC PERSPECTIVE

LET TALK ECONOMICS: DOES AUTONOMOUS DRIVING MAKE ECONOMIC SENSE?

Consumer (potential) attitudes to autonomous driving generally take two forms: the utopian view where the use of autonomous driving is heralded as ostensibly as good, with some of the advantages mentioned in earlier sections of this paper, including predictive models that expect driverless driving to save around 600,000 lives by 2045. In addition to this, driverless technology is also viewed positively as relates to the passenger, allowing them to use their new-found time whilst commuting to be more productive. The dystopian view leans more to the effect that these technologies may have on the more than 5 million truckers, cabbies and other drivers that depend on driving as a primary source of income.[1] It is more likely that the development of these technologies will bring with it a mixed bag: some positive economic benefits, as well as some negative. Finding the balance is key, especially if the technology can bring about a net positive outcome.

EMPLOYMENT - EXPLORING NET DIFFERENCES

A 2015 report by the US Department of Commerce, Economics and Statistics Administration (office of the Chief Economist) estimates around 15.5 million US workers that could, in one way or another (varying degrees), be affected by the introduction of automated vehicles. At the time, this figure represented around 1 in 9 American workers.[2] Of the 15.5 million, around 3.8 million are classified as 'motor vehicle operators', meaning that their occupations involved driving vehicles in order to transport people and goods as the primary activity. As a result, these workers were mostly concentrated around the transportation and warehousing sectors and can be considered to be at most risk from the threat that driverless vehicles pose on their livelihood. The other 11.7 million are classified under 'other on-the-job drivers', professionals that use roadway motor vehicles as part of their primary jobs that involve delivery of services or may need to travel to work sites. Such jobs may include first responders, repair and installation, personal home care aids and perhaps construction trades. Though they may not be part of those directly affected by the advent of driverless technology, these jobs are likely to be improved as it would allow for cheaper per-mile travel, as well as freeing up these service providers' time, allowing them to complete more administrative work whilst on route

1   Marshall, A. (2017, June 3). Robocars could add $ 7 trillion to the global economy. Retrieved from https://www.wired.com/2017/06/impact-of-autonomous-vehicles/

2   Beede, D., Powers, R., & Ingram, C. U.S. Department of Commerce, Economics and Statistics Administration. (2017). The employment impact of autonomous vehicles. Office of the Chief Economist.

to their customers.

A wholesale shift

One in 9 U.S. workers are considered on-the-job drivers. Of those, 3.8 million are motor vehicle operators and could be displaced by self-driving vehicles. Another 11.7 million use cars to deliver services or travel to work sites, and could see a boost in productivity.

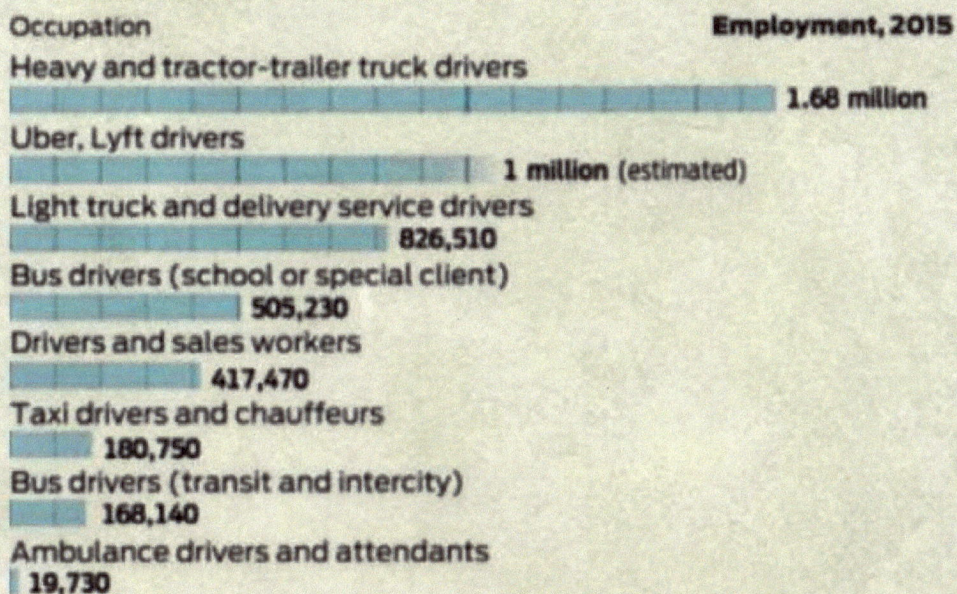

Occupation	Employment, 2015
Heavy and tractor-trailer truck drivers	1.68 million
Uber, Lyft drivers	1 million (estimated)
Light truck and delivery service drivers	826,510
Bus drivers (school or special client)	505,230
Drivers and sales workers	417,470
Taxi drivers and chauffeurs	180,750
Bus drivers (transit and intercity)	168,140
Ambulance drivers and attendants	19,730

Source: US Department of Commerce

JOB LOSSES ARE IMMINENT

A report by Goldman Sachs reveals the full extent to job loss in the American perspective. The report estimates that the American truck driver will be hit the most by the implementation of driverless technologies in everyday life. In fact, at its peak (according to the report), US motor vehicle operators are likely to see job loss rates that near 25,000 a month, or 300,000 a year.[3] Of the close to 4 million motor vehicle operators, around 75% of them were truck drivers, representing around 2% of the total level of employment in the US. These numbers assume that around 20% of all car sales will be constitute semi and fully autonomous car sales between the year 2025 and 2030.

Some of the social ramifications of job losses will be handled as part of our focus in later parts of this book. Some of these include the number of US drivers that work for ride-hailing services such as Uber and Lyft that, in the medium and long term, will likely be cannibalized by these driverless technologies, some of

3   Balakrishnan, A (2017, May 22) Self-driving cars could cost America's professional drivers up to 25,000 jobs a month, Goldman Sachs says. Retrieved from https://www.cnbc.com/2017/05/22/goldman-sachs-analysis-of-autonomous-vehicle-job-loss.html

which are ironically being developed by Uber.[4] Autonomous vehicles will also make increasing in-roads into other carrying tasks such as transporting chemicals within factories and laboratories. The following image shows automated vehicles carrying products.

Source: Breakawaystaffing

It raises an important question: where will those replaced by technology do to add to their livelihoods, and if the opportunities exist, what training and skills will they require in order to make the transition?

4   Engelbert, C, & Corwin, S (2017, August 1) Driverless cars and trucks don't mean mass unemployment – they mean new kinds of jobs. Retrieved from https://qz.com/1041603/driverless-cars-and-trucks-dont-mean-mass-unemployment-they-mean-new-kinds-of-jobs/

The Horse is here to stay but the automobile is only a novelty - a fad

Michigan Savings Bank
1903

A LOT OF THOUGHT IS NEEDED

The truck driver problem is not a small one, even when considering the positive impacts that may be presented by a driverless future. Net positive impacts are a good thing for the entire economy; however, there still remains the actually unemployed individuals, most of the likely ones being truck drivers. This is also true for all those industries that are considering 'supportive' of those that deal directly with fleets of vehicles. As discussed, the advantages of driverless technologies are significant, such as increased efficiencies in fleet management, significant savings in labour costs, including general savings, of all of which can be passed down to the final consumer, and that may very well be you.[1] Which is essential to the question: what will all these people without jobs do?

"The entire car ownership model is under threat, and with it, the wide range of jobs it currently supports: everything from repairs to parking, from financing to insurance."

~ Ernst & Young

So, you may ask, what are some of the positive outcomes that may be achieved by those companies specifically in the trucking industry, that specifically may warrant job losses in the trucking industry? For one, optimal fuel usage. Studies seem to agree that driverless technologies present fuel consumption savings upward of 10% compared to human drivers. These savings stem from cars (integrated with autonomous driving technology of course) that will be able to attain and maintain optimal cruising speeds through the use of computer-controlled acceleration and braking. Driverless technology is also expected to allow the optimization of follow-distance, meaning that vehicles will be able to maximise on the benefits of drafting, as well as building on the aerodynamic capabilities of vehicles.[2] In addition to this, and as mentioned in earlier sections, driverless technology is expected to reduce the number of crashes, largely caused directly by human error, by around 90%. Putting a price on human life is difficult, naturally, but this is undeniably a good thing. Other costs in the process are labour cost, which account for around 75% of all costs within the trucking industry, as well as increasing the number of hours that trucks can be used for given that machines are able to bypass restrictions that would normally apply for human. Such restrictions may include current federal laws that restrict the number of hours that truck drivers can drive

1   Ernst & Young. (n.d.) What will the drivers do when the cars are driverless? Retrieved from https://betterworkingworld.ey.com/workforce/the-future-for-drivers-in-the-driverless-future

2   Donley, J. (n.d.) Automation will change the trucking industry forever. Retrieved from http://aciesgroup.com/wp-content/uploads/2017/01/Automation-Will-Change-The-Trucking-Industry-Forever.pdf

consecutively for, as well as the minimum amount of time off that must be given between the shifts of individual drivers.[3] Essentially, trucks driven using autonomous technologies will be able to operate day or night without interruption.

However, as the title suggests, there is a lot to think about, meaning that although there are negatives foreseen, there still exist opportunities that may present themselves, as is expected of every period in which disruption has had its hand in. Some may include companies that upgrade to driverless technologies, such as those in the taxi or haulage sectors, may have to consider the implications brought about by such changes. Other include having conversations around who bares the responsibility of retraining and retooling of displaced drivers, especially since many companies are expected to see shifts in their labour force composition (more tech, less people). Other questions around the legal and regulatory environments are expected as well, such as how law makers and regulators are expected to adapt laws and norms for drivers and adapt the same for autonomous vehicles.[4] In addition to these, business that may consider switching from manual to automated labour may need to consider new elements of risk within their risk management, as well as have an acute understanding of the various legal obligations they would have to meet continuously. Additional risks to be considered for companies considering the switch may take the form of employment law, change in insurance coverage, as well as cyber security.

3   Ibid
4   Ernst & Young, ibid

THERE'S A LOT TO BE POSITIVE ABOUT

Though it makes for good reading and catchy headlines, it is not all gloom and doom that driverless technology promises. In fact, the opposite is true. Goldman Sachs strike a positive tone in their report, highlighting that many other models, including its own, do not show massive disruption caused by labour replacing technologies.[1] For one, it is expected that, as with many other technological advancements, new types of mobility (autonomous vehicles are included here) are likely to lead to gains in efficiency and productivity. This will result in making each mile of travel meaningfully less expensive.[2] The net effect of this is that consumers are likely to see a surplus income available to them to spend on other goods and services. For one, our current Baby Boomers (generation born in between 1946 - 1964 and make up around 20% of the American public) are likely to spend their additional money to pay for aides that may need to travel with them for medical appointments and such like services, even if driving may not be part of the overall job description. Other sectors that may see increases include travel and leisure.

Other areas of employment are likely to see rises in their absolute levels: lawyers, given the rise in legal complexity viewed as one of the probable ways in which employment may not necessarily always drop. For example, a report produced by KPMG for the Society of Motor Manufacturers and Traders (2015) suggests that driverless cars will likely accelerate the renaissance of Britain's automotive industry by creating hundreds of thousands of jobs, which is not only expected to save thousands of lives, but also add tens of billions to its economy.[3] The report suggests that Britain's continued embracing of autonomous cars and the technologies that are incorporated in them is likely to lead to the creation of around 320,000 jobs in the United Kingdom, 25,000 of which will be in the automotive manufacturing. It is also expected that around £51 billion (approximately US $69 billion) will be added to its economy each year, in addition to the technology preventing more than 25,000 serious accidents and thereby saving around 2,500 lives by 2030. The UK has been able to embrace the driverless vehicle because the country did not ratify a convention by the European Union that required vehicles to have a driver in them, meaning that the country did not require to new legislation in order to begin trials on its roads.[4]

As expected, the country has been able to use this loophole within its current

1.   Balakrishnan, ibid.

2.   Engelbert, ibid

3.   Tovey, A. (2015, March 26). Driverless cars to create 320,000 UK jobs and save 2,500 lives. Retrieved from https://www.telegraph.co.uk/finance/newsbysector/industry/engineering/11495706/Driverless-cars-to-create-320000-UK-jobs-and-save-2500-lives.html

4.   Tovey, ibid

legislative structure to champion its development of the technology, and this has led to the introduction of four pilot schemes (as at 2015) in London, Bristol, Milton Keynes and Coventry in an effort to discover how the technology can be incorporated on to Britain's roads. Further, in addition to the £19 million (approx. US $26 million) that the Government has put in funding for the schemes highlighted, Chancellor George Osborne announced an additional £100 million (approx. US $135 million) a sum that will be matched by players in the industry). This signifies the importance that Ministers place on the chance to grab pole position with regards to the implementation of the technology.

"New technology is fundamental to government's vision for our roads. Connected and autonomous cars will help us move to a smart, safe, efficient and low carbon future."

~ Robert Goodwill, Transport Minister, UK (2015)

In general, there is likelihood that the market around mobility managements services could either offer incremental job growth, or new jobs altogether. Other businesses that may crop up may offer digital planning and consumption of passenger and foods movement, making them more efficient, enjoyable, productive, safer, cleaner and cheaper. Such services may involve range from fleet maintenance to remote monitoring of said fleet, all of which would augment well with driverless technology. Just because the car can drive itself does not mean that it is fully autonomous in all respects, and there exist potential opportunities for industries to add value to the technology (thus improving the product as a whole) whilst taking advantage of the financial returns on offer.

There is more to life than simply increasing its speed."

Mahatma Gandhi

CONCLUSION

"In short, when autonomous vehicles take over the driving, drivers will need to retool their skill set to relevant opportunities that this new environment creates - akin to the industrial revolution or even the mass adoption of computers, the internet or even the mobile device."

~ Ernst & Young

From an economic standpoint, therefore, it would be difficult to ascertain the overall impact autonomous driving will have on employment, especially those jobs that are in direct competition with the technology. Industries that have driving at the core of their product or service delivery are likely to be affected rather significantly by the adoption of autonomous driving. In what is to many an obvious comparison to the industrial revolution, it is likely that job attrition is likely to hit the trucking industry hard, especially with regard to the number of jobs that are likely to be lost (or demoted) as a direct result of driverless technology. However, these companies are also likely to gain from savings to their current cost structures, where it is estimated that around a third of the cost of trucking is down to labour.[1] In addition to the reduction in operational costs, it is also expected that driverless technologies are likely to reduce the number of accidents on the roads. Further, it is difficult to estimate the number of jobs that will be created by other industries that are likely to be directly involved in driverless technology or provide support services. From an economic standpoint, it is more likely that there is likely to be a net positive effect when all factors are taken into consideration.

Another key consideration is the overall expected economic benefits brought on by autonomous driving technologies and their implementations. In the long haul, autonomous driving technologies are expected to contribute to around US $ 7 trillion in the US alone. Job losses are likely to be netted off by gains in other sectors of the economy. The net difference (overall impact of the technology) is more likely to be positive, as some of the expected gains from the technology are difficult to put a dollar value to, such as the preservation of life, total savings from accidents prevented, as well as the fact that any gains in productivity are likely to be established post-implementation of the technology. However, early indications from the experts' part of the automotive industry estimates that humanity is likely to benefit significantly from the development of these technologies. Though the economic benefits of these technologies are only likely to be achieved in the longer-term, especially when the adoption of the technology reaches optimal capacities.

---

1   Donley, ibid

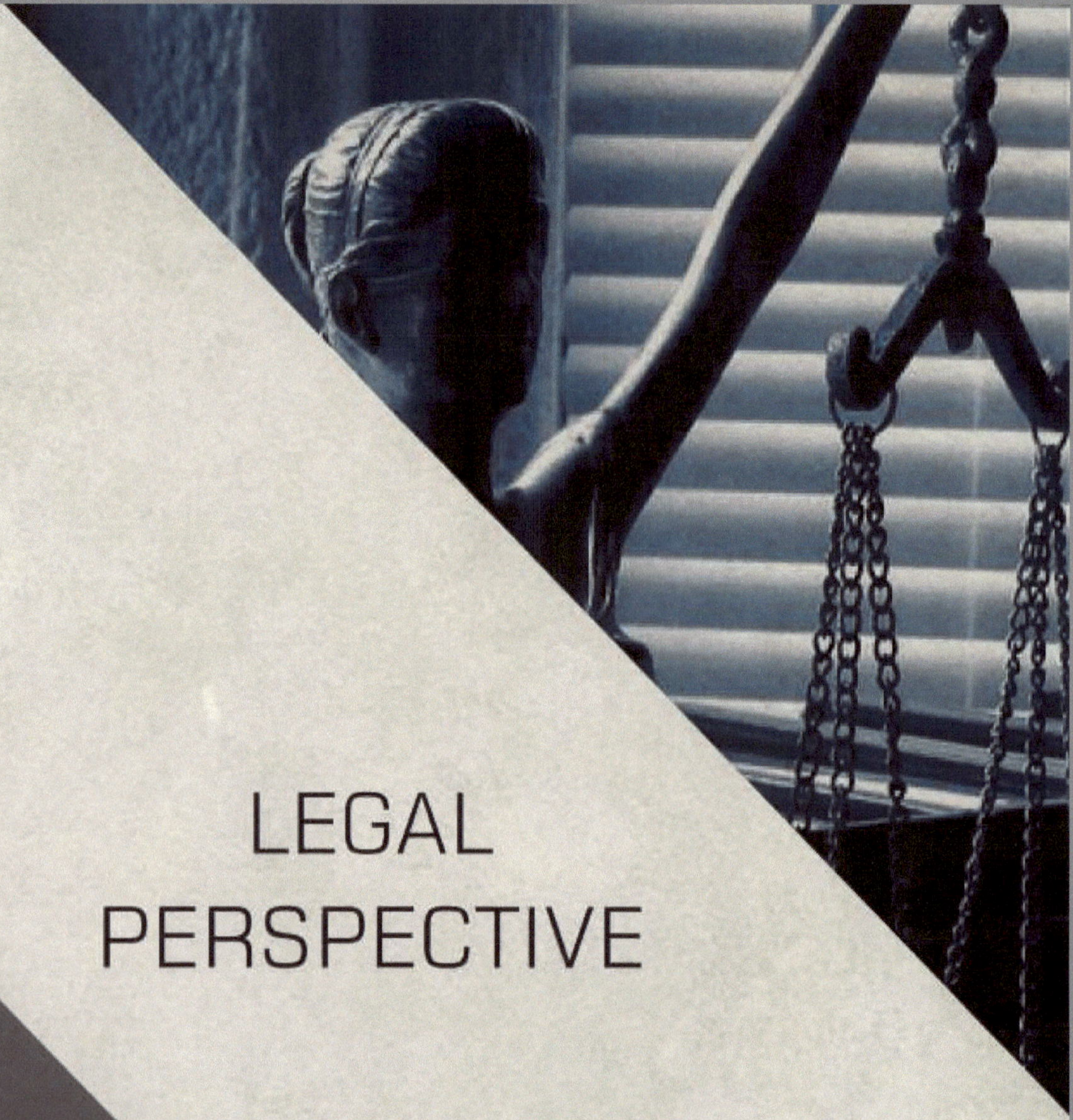

LEGAL PERSPECTIVE

CHAPTER 04

LEGAL HISTORY

Transportation has also been a high area of risk for human beings. Horse-related collisions were common in the pre-automobile times and the perils of the sea for ships have a long history of tragedies, loss and harm to persons and property and typically extensive legal issues at the end. The advent of the automobile also led to a dramatic rise in the number of court cases dealing with negligence and even manslaughter.

Over time the onus of dealing with dangers of automobiles has shift to governments who are expected to set and enforce detailed regulation on numerous matters involving car use, speed, seat belts, registration, driver's behaviours. Traffic enforcement concerns a significant about of police time. In the United Kingdom, for example, there are over 5,000 police officers dedicated to traffic offences alone. The situation is similar in the US. Behind the frontline law enforcement personnel there is also an army of insurance investigators, personal injury lawyers and various agents each making a living in one way or another from automobile accidents.

There are also other legal issues with implications to the transport industry. Compliance issues as well as anti-trust issues are particularly relevant to the sector.

WHAT COULD HAPPEN LEGALLY WHEN A PERSON IS INJURED IN AN AUTONOMOUS CAR?

There are numerous legal issues that arise out of car accidents involving a motor vehicle that are self-driven. An accident once occurred in Florida involving an autonomous car. The pressing issue is on who is responsible for the injuries suffered and ultimately who shoulders the legal responsibilities of paying the damages that accrued in the accident. These issues have left so many questions unaddressed on the legal position considering the principles that guide prosecution of cases involving an accident and the responsibilities thereto. For instance, who assumes the responsibility in a circumstance where the owner of Tesla driving in autopilot mode hit a truck? The issue of the insurer may also be raised in such scenario. For these legal questions, it is prudent to seek legal advice from legal experts with vast information on how to canvass these issues.

The term autonomous car is somehow misleading considering the legal issues at hand associated with the technology. For instance, if a fully autonomous car stuck another road user would the case proceed through the vehicle insurance in the normal manner or the party aggrieved will have to bring a potentially complex liability against the manufacturer?

The current autonomous cars are fitted with features designed to avoid an accident by alerting the driver of the potential harm or implement safeguards that take over control of the vehicle. The important insights on the evolution have a direct implication on the manner driverless vehicles are conceptualized. The current laws are based on the idea that humans are in control and should take over during critical circumstances. For better legal framework there is need to address fully the autonomous completely to a point where they perform verifiably better than all human drivers.

By the mid 1980s, automated autos, noiseless pneumatic subways and luxury-liner hover-craft will have radically restructured our surface mobility.

David Rorvik
1970

WHAT ARE THE SAFETY MEASURES PUT IN PLACE FOR THE AUTONOMOUS CARS?

It is with no doubt that these are issues that people will be confronted with as rapid advances seem to move towards that direction where autonomous car will dominate the roads. It seems in the near future it may be unnecessary to take someone to driving school due to the introduction of the autonomous cars.

The proponents of the autonomous cars suggest that it will improve safety in our roads and facilitate the efficiency in terms of ease of traffic and fuel consumption. The circulation of the autonomous car will be allowed in California without having a driver on board by the end of the year.

However, incorporating a new technology always takes time for the people to accept and adopt as their mode of operation. Majority of the people still have trust issue on the technology despite the assurance that the technology is particularly for the purpose of improving the safe use of the roads. This aspect seems to persuade the vast majority, and we may all end up adopting the autonomous car technology.

HOW DOES THE TECHNOLOGY IMPROVE ROAD SAFETY?

The case of Waymo has been used by Rocky mountain institute to evaluate on the conventional ways with the autonomous cars. Waymo has done a test drive for over 2 million miles within the United States. He recorded only a single accident making the accident risk 10 times lower that the conventional means that involves best drivers and 40 times lower than the danger facing the new drivers.

Additionally, owing to the autonomous cars, the risk and accident associated with drunk driving which represents considerable percentage generally will highly be eliminating by the autonomous cars.

WHAT ARE THE IMPACTS OF AN AUTONOMOUS CAR ON THE INSURANCE INDUSTRY?

Owing to the emergence of the autonomous vehicle being on the road, there will be significant impact on the insurance industry who are currently significant stakeholders in the automotive sector. The person responsible for an accident is no longer "a person," but rather a machine.

This aspect changes everything and the policy framework on how to handle the whole process in case there is an accident. Some of the insurance companies have welcomed the idea and are willing to collaborate in order to combat the situation. However, there will be significant and large-scale changes to the insurance industry. Most likely, legislative reform will be required.

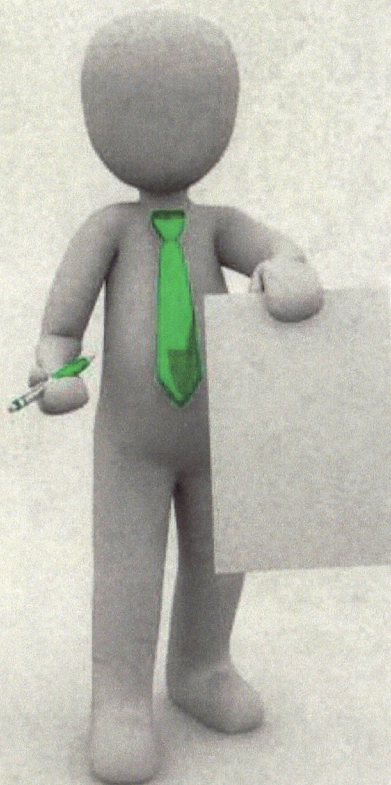

WHY IS THERE A NEED FOR LEGAL REFORM REGARDING ACCIDENTS INVOLVING AUTONOMOUS VEHICLES?

The fatal Tesla accident that killed Mr Joshua D. Brown exposed some of the fundamental shortcomings in autonomous driving, and the legal framework existing to deal with it.[1] The autonomous vehicle received heavy criticism specifically on the issue that there are human factors involved to supervise driving automation software and correct its flaws.

The accident raised ethical debates about autonomous cars. Additionally, the legal issue that needs to be answered or rather understood before pursuing the matter for judicial determination is who are the real parties for the cause of action in the accident, the type of the accident, the driver or the manufacturer or the programme developer.

It is important to answer these questions because when an AV is involved in an accident, there are multiple "parties" that will be involved. Who carries the blame the autonomous car or its driver or the manufacturer or the programmer who designed the cars performance guidelines. The debate seems to open a Pandora's Box on multiple other endless issues to be responded before just determination is considered by an adjudication officer.

1   Hars, A (2016, July 11) Fatal Tesla accident exposes fundamental flaws in the levels of driving automation framework. Retrieved from http://www.driverless-future.com/?cat=9

KEY QUESTIONS

Despite the positive impact of the autonomous vehicle, there are some of ethical implications that must be considered appropriately. Perhaps another issue to ponder on autonomous vehicles is that of whose safety the car will be programmed to protect in the event of an accident particularly with regards to vulnerable road users.

IS IT THE DRIVER'S FAULT?

Conventionally, the driver is the first accused person in an accident. Reason being that the driver is in full control and thus bears all the responsibilities in case something goes wrong. In the autonomous car, for instance in the case of the Tesla accident, the national transportation safety board claim that the deceased driver Joshua Brown was travelling at 74mph in an area where the speed limit is capped at 64 mph. It is noteworthy that the vehicle does not have the capacity to operate fully on its own. There is a fundamental difference between the continuous driving system and the short operation by the driver.

IS IT THE MANUFACTURER'S FAULT?

Evidence suggests that Brown deferred several rules that Tesla imposes on drivers. In essence, Tesla suggests the driver is at fault for ignoring the instruction manual of his car. It is indeed true that Brown ignored the instruction but Tesla cars are not classified as a vehicle for automation but for assisted driving. On the basis of this aspect, they are equally to blame as they had already assumed a co-driver role in the vehicle.

IS THE PROGRAMMING TO BE BLAMED?

If the responsibility is placed on the driver, it will be imperative to make consideration before making the agreement. If the fault lies with the manufacturer, then it will mean that if the burden shifts away from the driver then it will be the responsibility of the manufacturer to ensure that there is no mistake. But what if the fault is the programming of the car?

Developers will likely develop imperfect software if they can expect the driver to fill any gap. The Tesla case uses a non-redundant mono camera but what if the camera malfunctions or dies with the auto-pilot engaged and the driver fails to pay attention? What are the options considering the driver believes in the automation of the system?

Considering these issues that are likely to be borne of contention, we would enter into a litany of controversies that affect the autonomous car. On account of these issues, it is appropriate and wise for any person seeking an answer on the issue on the autonomous car to seek legal guidance from an attorney particularly with vast knowledge in matters accident involving autonomous cars. The debate on the ethical issues and the legal responsibility may not be an easy task to interrogate to a substantive conclusion without the assistance of a legal expert.

CONCLUSION

There are several moral and ethical considerations when considering the ethics behind autonomous vehicles. For one, the burden of responsibility problem (with respect to who bears responsibility for an accident) is very real. As discussed earlier part (and in truth, in later parts as well) of this text, the traditional automotive market has already been figured out; he who is found to be culpable for an accident will ultimately bear the repercussions for it to the extent of the damage caused by the accident. Therefore, if a person dies in an accident in relation to wanton and reckless driving of another, the driver in the wrong is expected to bear the legal responsibility for it. Any fines and forms of punishment that arise from the accident are expected to be borne by the individual that is deemed culpable, or the insurance company that has a policy with the individual, or both. The ethical and moral conundrum associated with driverless vehicles is that 'driver' is the software and vehicle hardware (technology), a fact that raises issues around whether the software and hardware companies can be reasonably expected to be held responsible when an accident occurs. As we will find out later, the accident involving an Uber vehicle fatally hitting a passenger 'chose' not to avoid the victim, even though the circumstances behind this choice are still undetermined.

Other ethical issue that arises whenever a driverless conversation takes place involves the ethics around job losses. Should we, as a human race, embrace technologies that take significant populations out of the job market? Are we ready to deal with the societal impacts of such actions? Perhaps the assumption underlying these arguments are sound and reasonable, or they may be ignorant. As earlier intimated, it is difficult to estimate the effects that these technologies are likely to have on society with respect to the job market. However, humanity has had several other instances in which the introduction of a 'disruptive' bit of technology caused mass concern, as it should. Take for instance the invention of the factory line of production that allowed mass production. Concerns early on at the time revolved around the safety of the manual jobs at the time. However, efficiencies achieved from the use of mass production spurred economies that used these technologies, and there is no reason to think that this would not be the case for the autonomous industry.

Finally, consider this: is it ethical to potentially miss out on adding US $ 7 trillion to the US economy, especially considering that this number is likely to lead to a net benefit to the economy despite the expected loss of jobs?

ETHICAL
PERSPECTIVE

ETHICAL PERSPECTIVE

Driverless technology, even when considering the overall positives brought about by it, has a few ethical considerations. Some, such as job losses, fall under both economic and ethical considerations. Others however, fall squarely within the ethical side of the debate. Issues such as amoral agency, that is, the moral decisions made by drivers, would have to be transferred to the driverless vehicle. Given that decision making is transferred to the vehicle, some hypothetical (but perhaps reasonable) examples of situations in which a moral problem to be solved by the technology may be considered, and the potential outcomes discussed.

THE TROLLEY PROBLEM

Chief among the ethical considerations to be made revolve around the trolley problem, a fairly popular thought experiment in ethics. In its general form, the problem suggests the following set of circumstances:

> There is a runaway trolley (carriage as part of a train) barrelling down a railway track. Further down the track, there are 5 people on it, tied up and unable to move away from the danger that it poses to them. As you stand in the distance, you have access to a lever that, if pulled, will switch the trolley to a subsequent track, meaning that the trolley ceases to pose a danger to the 5 people on it. However, in so doing, you notice the trolley would be diverted on to a track that has one person tied up and unable to move. In such a scenario, you would have two options: a) do nothing and the trolley will most certainly kill the 5 people on the track; or b) pull the lever and the one person on the alternative track is most certainly going to die.

The problem with the trolley problem is the moral agency that it applies: on one hand, if you do nothing, you will have had no hand in the fate of the individuals on the track. The trolley problem exists within several variants but poses the same fundamental question: would you save the lives of the 5 individuals, but in so doing, play a direct hand in the death of the one?

This type of problem poses significant questions to be answered by autonomous technologies. Such vehicles will not only be tasked with making the decision of whether to kill one or five, but will also be tasked with answering questions around whether they will be optimised for overall human welfare, or would they prioritise the passenger safety? Perhaps those on the road would be prioritised? The moral questions to be answered by these technologies number in the tens at the very least, with the waters muddied even more so once the consumer is added to the fold, making things potentially thicker.[1]

1   Shashkevich, A. (2017, May 22). Stanford scholars, researchers discuss key ethical questions self-driving cars present. Retrieved from https://news.stanford.edu/2017/05/22/stanford-scholars-researchers-discuss-key-ethical-questions-self-driving-cars-present/

ASSESSING AND MINIMISING RISK LEVELS

The Centre for Automotive Research at Stanford (CARS) argues that agonizing over the solution to the trolley problem is beside the point. Whilst representing the centre as executive director, Stephen Zoepf suggests that it is not productive to assess whether driverless technologies would be able to solve complex problems, but whether the technology would, in and of itself, be able to earn the trust of the society.

"It's not productive. People make all sorts of bad decisions. If there is a way to improve on that with driverless cars, why wouldn't we?"

~ Stephen Zoepf, executive director,

Centre for Automotive Research at Stanford (CARS)

The challenge, therefore, for driverless technologies is whether or not it can make a social case for its existence within society. This means that developers of this type of technologies would need to ascertain the level of risk that society in general would be willing to incur with said cars. In so doing, developers would have to examine the trade-offs inherent in safety and mobility, that is, the sacrifices that customers would have to contend with in. Some of these include whether autonomous vehicles would be required to have right of way (perhaps to minimise the risk of accidents by reducing exposure of the technologies to human elements), or the speeds that society would be comfortable permitting.

Let's prioritize ethical considerations at the front end of design

John C. Havens

IT WILL BE DIFFICULT TO DEVELOP ETHICAL ASSUMPTIONS

Making a social case for driverless technology rests on its programming, which is most likely to be developed partly in conjunction with local governments. Local governments are likely to want to ensure that the technology maintains society at its current state, and if possible, improve on it, whilst also providing it with advanced services. This means that whilst autonomous vehicles would be providing utility in the way of savings for governments and citizens alike, the technology must also ensure that it does not pose significant challenges to the same society.

Some of the challenges that autonomous vehicles may face within the ethical dimension is that of how to write into their algorithms code that will govern the initial set of fully autonomous vehicles. The technology behind these vehicles works by constantly (and at rather quickly) collecting information from its environment with the help of sophisticated cameras and sensors that rely on technologies such as laser-based ranging ('lidar'), ultrasound and radar. Such technologies, as well as a host of others, enable autonomous vehicles to correct for human mistakes, and thus allowing them to 'learn' from the collective 'experiences' from all other autonomous vehicles within a specific network. These autonomous vehicles will store data that will be used by engineers to reconstruct typical accidents, and in doing so, allow them to analyse the multiple inputs that the car sensed using the aforementioned technologies. Not only that, the engineers would also be able to assess the logic that the vehicle used in order to determine the course it took during the crash.[1]

Engineers would not be the only individuals required in this process. Manufacturers and software designers will be able build on the work done by engineers in order to develop modified programs that can use this information to make better future decisions, thereby improving the vehicle overall. However, with every improvement (iterations), the vehicle will have to be based on sometimes vastly different sets of ethical assumptions, meaning that science will have to explore moral behaviour (some would call it 'ethical crashing algorithms') for autonomous vehicles (AVs). As discussed, an ethical question to consider could be "who to kill in the event of a crash?" Two pedestrians or 2 autonomous vehicle passengers? Autonomous vehicles are bound to find themselves in situations such as these fairly routinely, a move that will likely lead to the need for underlying ethical assumptions that the car would use in order to chart a course of action. Therefore, forced choices such as those highlighted above and in other sections of this text may be required to be programmed in conjunction with sophisticated algorithms

1   Fleetwood, J. (2017) Public health, ethics and autonomous vehicles. American Public Health Association, 107(4), 532-537

that would rest on fundamental ethical assumptions that are not yet articulated (or would need improvement).

"...autonomous vehicles may prove to be the greatest personal transportation revolution since the popularization of the personal automobile nearly a century ago".

US Department of Transportation: National Highway Traffic Safety Administration

FORCED CHOICE ALGORITHMS

Consider that autonomous vehicles will not only be required to follow the rules on the road, but they may also be required to, within certain circumstances (naturally), to make ethical decisions. Current day drivers (humans) are frequently, as part of driving in the real world, required to make sophisticated, nearly instantaneous ethical decisions. These decisions come as a result of ethically challenging situations that drivers are put in as part of their daily commute, and it would thus be overly simplistic to think that AVs would only need to follow the rules of the road in order for them to be considered safe. Some instances in which such a vehicle would be required to perhaps look past a traffic rule due to the ethical implications behind said rule would be when a vehicle would need to cross over a double-line lane, and on to oncoming traffic, so as to avoid hitting a vehicle on the shoulder of the road. Others may involve drivers crossing a red-light in order to get out of the path of an oncoming train.

> *"...automated vehicles must decide quickly, with incomplete information, in situations that programmers often will not have considered, using ethics that must be encoded all too literally in software. Fortunately, the public doesn't expect superhuman wisdom but rather a rational justification for a vehicle's actions that considers the ethical implications. A solution doesn't need to be perfect, but it should be thoughtful and defensible".*

~ University of Virginia Transportation Research Council

Some ethical dilemmas emerge when considering the realm of forced choice, or to put it another way, having to make a decision between two bad choices. Similar to the trolley problem it is interesting to note that a small study on autonomous vehicles yielded that respondents preferred others to make utilitarian choices whilst they themselves would prefer preferential choices. The study sought to establish how participants would react to several hypothetical forced-choice accident scenarios and were asked to choose between whether pedestrians or passengers should be given priority if the situation would have to involve the death of one of the parties. The study found that around 76% thought that the autonomous vehicle would be better placed and thus more utilitarian, in their view, if it were to sacrifice its own passengers in order to save more lives overall. However, as irony would have it, the same respondents to the study, when considering whether they would purchase autonomous vehicle themselves, gave more priority to vehicles that would preserve them or their family over the same utilitarian aspects. The respondents chose for the vehicles to protect their own lives over those of others, regardless of whether this outcome was the most ideal given the circumstances.

Overall, forced choice algorithms are likely to feature prominently when discussing the ethics around autonomous driving, especially the ethical implications that these algorithms will have on the overall conversation around autonomous driving. This is especially true when considering that there are significant inconsistencies in participants ethical reasoning (utility vs self-preservation), as well as the plethora of divergent views on the topic. It is therefore a necessity for engineers, manufacturers and software developers to consider the ethical challenges posed by forced-choice algorithms.[1]

1   Fleetwood, ibid

SELF-PRESERVATION IS A NATURAL REACTION

It is easy to understand - even though these analyses may be characterized as informed guesses at this point - why the general public may be in favour of self-preservation as part of the forced choice algorithms that would underpin driverless technologies. Drivers (let us call them primary passengers) would be presumed to be at greater comfort inside a driverless vehicle if they knew that the vehicle is programmed to place higher priority on preserving the lives of its occupants over any other individuals considered external to the vehicle, especially in the event of an accident. In the same vein, occupancy within autonomous vehicles that place higher priority of preservation on pedestrians, a situation that is likely to see would be primary passengers shying away from these technologies and instead choosing to take their chances with a live driver. Given that driverless technologies are likely to be adopted on the back of the prevention of human lives in general through the reduction of the number of road accidents, slow or reduced adoption of driverless technologies may then be viewed as an opportunity cost to that preservation.

Similarly, businesses that would like to position themselves as part of the mobility services sector, such as taxicabs, may prefer to have their vehicles to place preference of preservation on its occupants should the vehicle be forced into a forced choice situation. The idyllic situation, it is argued, is the converse of this, with all vehicles in a society programmed to give the other vehicle priority within any given forced choice situation, which is likely to lead to a net positive situation. However, for this to work, most if not all of the vehicles within a community would have to autonomous in nature, and in addition to this, the general public within said society would have to find such a forced choice algorithm acceptable to them. As we have discovered, the forced choice algorithm that prioritises occupants within a vehicle may not necessarily work with the larger consumer based, naturally.

THE UBER CRASH

A 49-year-old woman was hit and killed by an autonomous Uber on 19 March 2018. Though the specifics of the crash are currently under investigation, some details emerged from the crash. Elaine Herzberg, the victim of the crash, was pushing a bicycle loaded with packages at around 10pm at night in Tempe, Arizona. She was hit whilst trying to cross a brick pathway from a centre median that separated the two opposing lanes of traffic. Interestingly enough, however, is that she was hit whilst trying to cross a brick pathway that may have been inviting to cross over, even though it did come with a sign that warned pedestrians against using it. It is also interesting to note that the accident occurred a fair distance away from the nearest crosswalk, the vehicle was moving at 38mph in a 35-mph zone, and that neither the car's safety system nor its safety driver made any attempt to brake.

> *"It's very clear it would have been difficult to avoid this collision in any kind of mode"*
>
> ~ **Sylvia Moir, Tempe Police Chief**

The Uber crash was tragic, as those covering the incident all agree. The ethical arguments/ questions in relation to autonomous technologies are numerous. For one, should we expect more of autonomous vehicles than we would of humans in the same capacity? This would not be an obscene question especially considering the increased safety that autonomous vehicles are heralded to be at the forefront of delivering. In addition, given the capabilities of the vehicles with respect to sensors and computing power that far outstrip those of a human driver, wouldn't it be reasonable to expect that autonomous systems should do better than human beings? Again, this may not be an unreasonable question to put forth to those developing the technology or writing the law around this sector.

Ultimately, autonomous vehicles will be required to fulfil the promise that this technology holds, meaning that increased scrutiny over shortcomings said technology may be completely justified. After all, it is difficult to place an acceptable loss rate when human lives are involved. The short of it, however, is that autonomous technologies will have to perform at close to the levels of their expected benefits, meaning that they will have to be as safe as they claim to be, even though measuring this may be difficult. And this is the crux of the matter: autonomous vehicle manufacturers, due to the increased importance they placed on safety, may have to contend with raised public expectations. This would inadvertently mean that any accidents or deaths caused by these vehicles will be viewed with more scrutiny and increasingly negatively, which may explain the controversy behind this crash. Well, that and the fact that this technology has yet to be rolled out.

It is also interesting to note that there was a human in the car tasked with monitoring the car's activity, meaning that the blame for this crash cannot be placed squarely at the feet of the technology. It would take a vehicle with level 4 or 5 autonomy (empty of course) to test the ethics behind autonomous technologies. In the Uber case (the investigation is ongoing and thus not concluded), it would seem likely the pedestrian hit during the accident should share some of the blame for this considering that they did not use a crosswalk.

CONCLUSION

There may be quite a long way to go with regard to figuring out the various ethical implications brought about by driverless technologies. Some of those highlighted within this text include joblessness, and the fact that many people, both directly and indirectly, would be put at a disadvantage by the wide-spread adoption of the technology. And this is even before one considers the various intricacies brought about by the sensitive nature of forced choice algorithms. It would be, however, remiss to consider only the negative ethical impacts of said technologies, as may be explained by human instinct. It is prudent to consider the net human loss from not adopting such technologies, especially when increased economic benefit is also expected.

Manufacturers of such technologies argue that their vehicles may never require forced choice situations due to the technologies that they employ, adding that these technologies would allow the car to pre-empt situations far ahead of time, thereby avoiding such situations. However, it is clear that autonomous technologies pose moral questions, especially considering the net positive effects.

ENVIRONMENTAL PERSPECTIVE

CLOSEST TO HOME?

The environment is perhaps one of the few things that binds many populations on the planet, as often witnessed by the devastation of vast swathes of land and territories by tropical storms such as Hurricane Irma recently. Whilst the debate around human caused climate change continues to rage, it is clear that many governments across the world are concerned about carbon emissions and its potential to drastically change our environment. It is expected that the environment would be a critical part of the conversation around driverless technologies. For context, the US has committed to reducing its greenhouse gas emissions by between 26-28% below levels recorded in 2005, a target that is to be achieved by the year 2025. In order to meet said targets, the US will have to ensure that it cuts emissions from the transportation sector that is estimated to contribute approximately 26% of total greenhouse gas emissions (2014). Light duty cars, such as passenger cars, contributed to 61% of total transport sector emissions.[1] However, there are several other considerations when looking into the potential environmental impacts associated with driverless technologies, such as the impact that such technologies have on urban planning. Interestingly but unsurprisingly, driverless technologies have not yet been figured out fully yet, and as such, cannot reliably estimate its impact on the environment. However, this section seeks to discuss some of the aspects of environmental issues associated with driverless technologies, as well as some commentary on how these issues are likely to impact or be impacted by the adoption of such technologies.

1   Kearns, M.A., Peterson, M., & Cassady, A. (2016). The impact of vehicle automation on carbon emissions. Retrieved from http://www.ourenergypolicy.org/wp-content/uploads/2016/11/AutonomousVehicles-report.pdf

INCREASED MILES TRAVELLED RISK

As discussed in earlier sections of this book, automation is likely to free up the time and concentration spent by drivers on the road every day. To this effect, driverless technologies will lower the opportunity cost of driving, that is, the reduced amount of input in to the process versus its utility will make driving more attractive, meaning that more people will use vehicles. The added ability of a driver to do more than one thing whilst on a drive is likely to encourage more people to take more trips, a situation that is likely to increase the vehicle miles travelled (VMT). The VMT is an aggregate of the number of miles driven by vehicles in a given year.

Consider also that user of automated technologies may prefer to live further from their places of work if it meant that they could spend their time doing things that do not involve getting behind the wheel. Further, people may be encouraged to take trips over staying at home if it meant that they did not have to focus on the driving part of the travel. People that live in densely populated cities with expensive parking may prefer to either send their vehicle back home or have it drive around the area.

Not only are driverless technologies expected to lower the opportunity cost of driving, it is also expected that certain demographics that were previously limited in their ability to use vehicles would most likely no longer be inhibited any longer. Such demographics may include people with disabilities, the young and the elderly. All in all, it would seem most likely that autonomous technologies are likely to increase VMT. An increase in VMT is not good for the environment in many respects. For one, it leads to a higher number of vehicles on the roads, products that are not necessarily good for the environment themselves. An increase in the number of vehicles on the road would mean that global populations would have to contend with increased carbon emissions from combustion engines.

RIDE AND CAR-SHARING SERVICES

All this could be mitigated, however, through the use of ride and car sharing models of transportation. As discussed, mobility services are likely to form a significant portion of sales generated by automotive manufacturers. Assuming that this is so when driverless technologies are adopted as part of mainstream knowledge and use, current car ownership models might cease to exist. Consider that autonomous technologies take away the human component in a vehicle, meaning that the importance of a personal vehicle to individuals is likely to drop. This is especially true when considering that autonomous vehicles are much more likely to be able to plan efficient routes as well as transport said people from point to point. Such a concept would mean that a single vehicle, through the use of driverless technologies, would be capable of transporting several clients at the same time (ride sharing), and if not, then service one client after another.

It is also interesting to note that this ties in with predictions made by McKinsey & Company estimating that the growth in the number of vehicle units sold is likely to drop, and the mobility services sector is likely to experience a surge in its earnings. Inadvertently however, such technologies may actually increase VMT if the total amount of the vehicles are used inefficiently, for example, if they make frequent passenger-less trips on the way to pick up other clients. However, as with most things associated with technology, software can be used in order to rectify such a scenario, by planning the most efficient routes.

CONGESTION PROBLEMS ARE NOT LIKELY TO BE SOLVED IN THE SHORT-TERM

One of the immediate challenges that autonomous technologies may not be able to address in the short term is the congestion associated with an increased number of vehicles added to the roads (both autonomous and traditional). Therefore, whilst autonomous vehicles are added to the road, they would not offer any reprieve in as far as congestion is concerned as there would need to be a significant amount of such vehicles on the road for the benefits to be achieved. Therefore, because it would likely take time for autonomous vehicles to outpace their traditional counterparts both in new car sales and total volume, autonomous vehicles are likely to contribute to the congestion challenges of the future, rather than alleviate them. An increase in congestion is likely to lead to several issues that may be of concern as it relates to environmental matters. For one, an increase in congestion leads to a drop in efficiency. One of the primary selling points of autonomous vehicles is that they would be able to operate at higher rates of efficiency when compared to their traditional counterparts. This high level of efficiency is touted as promising to improve fuel cost (whether electric or fossil fuel driven), and thus potential savings to both domestic and industrial users of such technologies (such as trucking businesses). Lower rates of efficiency are also likely to make the vehicles unattractive when compared to traditional vehicles that people already understand.

Another challenge that congestion may bring is an uptick in carbon emissions. It is important to remember that autonomous vehicles are most probably going to be introduced into society together with traditional vehicles, that is to say, autonomous vehicles are not likely to be sold exclusively in the short-term. Therefore, an increase in congestion is likely to increase carbon emissions in a sector that is already a significant contributor to the global carbon footprint. The silver lining however, is that autonomous vehicles could help with congestion in the short term by avoiding accidents as well as through the use of the most efficient routes of travel. More so, AV technologies are expected to avoid inefficient driving conditions by minimising on inefficient and abrupt stops and starts. Avoiding this kind of driving would mean that cars would be able to drive more closely together; more cars would therefore be able to fit within the same amount of road space, the net effect of which would be a reduction in congestion as well as a more streamlined driving process. This would be made even more possible by the vehicle-to-vehicle (V2V) communications that would be made possible on level 3 AVs and up. V2V would allow vehicles to drive closer together even at highway speeds made possible by simultaneous braking.

Overall, it is difficult to estimate the net effect that AVs will have on congestion, especially in the short-term. However, large scale adoption of AVs is most likely to alleviate traffic jams and general congestion within urban areas. V2V technologies would allow cars to talk to each other, essentially alerting each other of potential bottle necks and re-routing vehicles to avoid snarl ups. Add that to large scale adoption of AVs (at a point where AVs will have saturated the market) will significantly reduce congestion on roads.[1]

1   Kearns et. al. ibid.

FUEL CONSUMPTION

Initial studies on the effects of AV technologies on fuel consumption patterns, though speculative, seem to suggest that they are likely to reduce overall fuel consumption. Though this relies heavily on the overall penetration of AVs within society; the benefits of AV technology are closely tied to penetration due to V2V communications that allow for simultaneous braking and accelerating by multiple vehicles. As earlier discussed, the smoothening of driving (braking and accelerating) combined with other technologies such as adaptive cruise control and traffic-smoothing algorithms.[1] The impact of these refinements to the smoother driving characteristics of vehicles would be significant enough to yield a lower consumption in fuel even if VMT actually increases due to an adoption of vehicles by those that would not have ordinarily driven. These studies suggest that the smoothing of traffic flows and the minimizing of braking on highways could contribute to the reduction of fuel consumption by as much as between 23 - 39%. Other studies suggest that a 20% reduction in the accelerations and decelerations would lead to a 5% reduction in the amount of fuel consumed.

Again, it is important to stress that these figures are speculative; however, they mainly seem to point to overall reductions in fuel consumption, with variances arising from the quantum involved. Further, these estimates seem to rely heavily on the efficiencies that will be achieved with the adoption of algorithms that are expected to increase the amount of fuel saved (especially in situations with a lot of stop-and-go traffic. However, it is also important to note that these eco-driving practices, primarily from slower speeds and gentler accelerations by AVs, would likely lead to inefficiencies in congestion, and this would then cause increases in the amount of fuel consumed. All in all, though it is difficult to estimate, AV technologies are expected to provide at least a little system-wide benefit when considered against smoother driving.

Vehicle platooning, the method that refers to reducing the distance between vehicles (automatically) as well as other technologies are likely to save the European region close to 700 million litres each year and prevent more than 1.7 million tons of carbon dioxide emissions. And these types of savings are estimated to be achieved if all the vehicles in the European Union actively used adaptive cruise control, with significantly higher savings within the region with the adoption of level 4 and 5 AV technologies. Other studies by Volvo (Swedish car maker) found that their use of vehicle-to-vehicle platooning yielded a 10% improvement in fuel consumption. A separate study by the Intelligent Transportation Society of Ameri-

1   Worland, J. (2016, September 8) Self-driving cars could help save the environment – or ruin it. It depends on us. Retrieved at http://time.com/4476614/self-driving-cars-environment/

ca predicts that the use of 'intelligent transportation systems' (level 2 & 3) such as adaptive cruise control and wireless consumption would yield a 2 - 3% reduction in oil consumption as these technologies spread through the automobile market.[2]

WILL HIGHER HIGHWAY SPEEDS INCREASE VEHICLE INTENSITY, THUS LOWER YIELDS IN FUEL CONSUMPTION?

AV technologies will allow vehicles to travel at higher speeds as a direct result of adaptive cruise control and V2V communication that allows for platooning. However, some analysts provide a word of caution, suggesting that AV technology could increase the energy intensity of vehicles on roads due to the technologies discussed above, technologies that would encourage vehicles to move at faster speeds. These faster speeds are likely to increase the amount of energy that these vehicles consume overall (especially since the energy consumed in this case is derived from the combustion of fossil fuels) which would be damning for AV technologies considering they have been heralded as potentially very good for the environment. It is important to consider that highway travel accounts for between 33 - 55% of all road travel, meaning that an increase in highway speeds may increase the energy intensity of all light-duty vehicles by between 7 - 22%.

Despite concerns around the increase of energy intensity by these vehicles, a reduction or replacement of fossil fuelled vehicles with electric vehicles would lead to overall reduction in fuel consumption. In addition, AV manufacturers may be able to encourage the adoption and spread of electric vehicles by going around one of the most significant challenges posed, that is, the need to constantly charge them, or rather, the lack of a robust infrastructure around charging stations. The increase in the number of available charging stations as well as the adoption of fully autonomous technologies would mean that electric AVs would be able to charge themselves. Further, it has been noted that the use of shared ride and AV technologies together would mean that vehicles would make multiple runs in a day, thus reducing the need to warm the engines of fuel driven vehicles (catalytic converters found in these vehicles need to be warmed up from a cold start, which increases emissions). According to some automakers, fuel savings would be achieved even with the wide-scale adoption of low-level AV technology.

2   Kearns, ibid.

Hillars Aerial Sedan will be your flying car in 1967

Popular Mechanics Magazine
July, 1957

CONCLUSION

As have many some of the chapters concluded, there is a lot to think about when it comes to the development of AV technologies. This area shows a lot of promise in improving our lives from a social stance, which is most certainly welcome. Stakeholders would be best served if they considered some of the nuanced approaches to ensuring that AVs are best utilised within our society. Think of its a 'prevention rather than cure' approach, wherein the developers would consider the following:

Smart growth - studies on the impacts of AV technologies should be performed against all relevant environmental programs within a specific geographical region. Smart growth involves considering a wide range of conservation strategies that are intended to help protect human health as well as the environment, all whilst making communities more attractive and socially diverse. This type of growth model also promotes the efficient and sustainable land development through the optimal use of existing infrastructure as well as optimizing the footprint of the developed land, thus reducing wasted infrastructure and land.[1]

Some of the considerations that a smart growth model would involve the impacts that AV technology would have on public transportation, local street systems and parking requirements. The use of smart growth would sensitise city planners to the importance of brownfield redevelopment (clean-up of contaminated land for reuse), as well as trying to avoid the potential adverse effects on the environment that an increase in miles travelled scenario may present. Candid conversations and collaborations around these realities may see the advancement of the technologies even faster as there would be less pushback from the relevant authorities (because these protocols would be developed with input from the authorities to begin with). Local authorities and officials would also be best advised to anticipate and allocate for the effects that large fleets of autonomous vehicle sharing services may have on local communities. Other institutions that may also have to consider some smart growth initiatives would be municipalities; a decrease in the amount of parking required means that they would need to rethink parking requirements as relates to commercial and residential developments in local zoning ordinances. This would work towards making use of land efficiently.

Urban brownfield redevelopment: AVs has the potential to stimulate urban sprawl due to the added convenience that the technology brings, allowing people to live further from the city, in turn driving down the demand for brownfield redevel-

1   Harrington, A., & Schenck, S. (2017, July). Steering the environmental impact of driverless cars. Retrieved from http://www.gklaw.com/Godfrey-Kahn/Full-PDFs/ Steeringtheenvironmentalimpactsofdriverlesscars.pdf

opment. Surface parking lots and parking structures are often considered contaminated. The need for these types of structures that would already be existing would lead to denser city centres with additional green spaces, adding to societal value.

Tax incentives and renewable energy policies: governments of the world will have to consider how tax incentives may be able to aid in the development and roll out of AV technologies, assuming that they would have considered the merits and demerits of such a roll out. Would these governments offer different incentives or penalties for different combinations of AV technologies on vehicles, such as incentivising electric vehicles more than fuel driven ones? Considerations must also be made around the impact that AV technologies will have on the creation of new industries and what those industries would have on current models of transportations, such as the impact of AV fleets available on demand. All in all, governments would want to ensure that they have incentives and policies in place that maximise on potential utility from AVs to ensure maximisation of value creation.

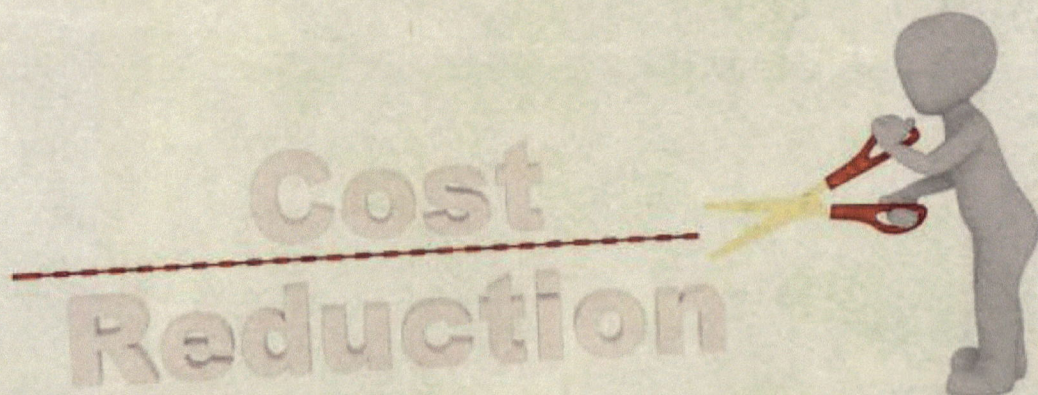

POLITICAL
PERSPECTIVE

THE FUTURE IS AUTONOMOUS? THE POLITICAL ANGLE

As discussed earlier, iterative improvements to driver assisted technologies are classified within autonomous driving, albeit between level 1 - 2. Such types of features are already widely available within current models of vehicles. Some of the technologies that have been part of cars for decades include anti-lock brakes, lane departure warning systems, rear view alarm systems, and adaptive cruise control.[1] Given the wide range of development paths available for autonomous vehicle technology deployment (incremental - automatic braking systems; larger - automated crash avoidance safety systems; and revolutionary - fully autonomous vehicles) when compared to existing systems, there are also likely to be different policy implications and regulatory interventions needed. Simply put, different levels of policies and regulations will be needed, depending on the level of autonomy that a vehicle is intended to perform.

SAFE ENOUGH?

The US Congress, as at late 2017, was in the process of considering legislation that would allow autonomous vehicle (AV) makers to deploy their testing vehicles within real-life settings as long as they were deemed to be as safe as current vehicles.[2] Given that we have established that the human being is the most unreliable part of the driving function, it would seem that autonomous vehicles would not struggle to match the levels of safety by these unreliable drivers. However, given that the US experiences one fatal crash for every 3.3 million hours of travel, matching current vehicles would not necessarily constitute your average walk in the park. Add to this the fact that human beings are much more likely to be forgiving towards human error when assessing a crash, much more than they would any crashes that came as a result of software glitches or malfunctioning hardware.[3]

Gill Pratt, CEO of the Toyota Research Institute based in California, believes that halving the current rate of road fatalities would not be enough to convince the general public that AVs are a safe source of transport. He suggests that policy makers would have to determine what constitutes a sufficient level of safety. However, some analysts, including Nidhi Kalra (senior information scientist at RAND Corporation) believe that iterative developments deployed, even if imperfect at the

1   Schreurs, M. A., & Steuwer, S. D. (2016). Autonomous driving - political, legal, social and sustainability dimensions. Autonomous Driving, pp. 149 - 171

2   Mervis, J. (2017, December 14) Are we going too fast on driverless cars? Retrieved from http://www.sciencemag.org/news/2017/12/are-we-going-too-fast-driverless-cars

3   Mervis, ibid

times, would be key to winning over public sentiment, as well as speed up the rate of improvements.

> *"Waiting for the cars to perform flawlessly is a clear example of the perfect being the enemy of the good"*

THE ARGUMENT FOR ITERATIVE IMPROVEMENTS

Nidhi, in conjunction with her colleague at RAND David Groves, co-authored a recent study that concluded by urging the US government to allow AVs on their roads as soon as they are able to achieve a 10% reduction in the number of fatalities when compared to current figures.[4] They argue that by allowing AVs on the roads, their self-driving algorithms would be better placed at learning from driving in the real world, rather than learning from trips around a test track or through the use of computer simulations. Further, they suggest that the use of imperfect AV by 2020 would go on to save around twice as many lives by 2070, as if the government would have introduced the same in the year 2040.

It is interesting to note that the study does not suggest that AVs will be perfect at a go, and thus, might they themselves contribute to accidents. However, the argument is that deploying the AVs early would help the industry overcome the narrative around AVs - they must never do harm.

> *"They will never be perfect. And it's going to take decades until they are 10 times safer."*

4   Ibid.

LET'S TALK ABOUT POSITIVE AND NEGATIVE EXTERNALITIES

Let's start simple. We can all agree that we all require our policy makers to write laws that benefit the general populace, mostly through policies that stimulate the economy of any given region, thus leading to more jobs and better general standards of living. This is critical to good policy making. In this light, consider positive and negative externalities (consequences of an economic activity), and their effect on the balance in markets.

Policy makers have to ensure they do not introduce any policies that may cause any market failure - a situation in which the equilibrium level in a market is flawed and is caused by significant externalities. These externalities incentivise the system to reward individuals at the expense of the group benefit.[1] Simply put, policy makers want to ensure that they do not introduce laws that may hand benefits to individuals, rather than the entire group. A good example would be a company that produces cars, but at the cost of polluting the environment, meaning that the company does not account for all the costs of their product within its overall costs, and is thus borne by the society that has to deal with the pollution effects. In this case, the car company would be accruing more benefits because the cost of their cars does not include pollution costs. If it did, it would be priced higher and the company would reduce their pollution. Therefore, part of ensuring that policy will not cause a market failure is that they would have to ensure that the benefits of autonomous driving are realised by all stakeholders. Autonomous driving must present benefits to cities, individuals and the environment, or at least ensure that externalities are internalized to the greatest extent possible.[2] There are a number of sources online that deal further with positive and negatives externalities in general.

1   Investopedia. (2018, February 13) How do externalities affect equilibrium and create market failure? Retrieved from https://www.investopedia.com/ask/answers/051515/how-do-externalities-affect-equilibrium-and-create-market-failure.asp

2   Ibid.

OVER-REGULATION: THE NEED FOR CENTRALISED POLICIES

The development of policies at a state level, even though it may be good in capturing the essence of the needs of the state, may lead to situations in which multiple states have conflicting laws that make it difficult to deploy a single autonomous system across multiple states.[1] There would need to be a centralised system that would be used to evaluate the compatibility of an autonomous system within specified parameters before it is allowed to exist within the general public. In a similar way to internet security, there is a strong justification for an international set of rules concerning autonomous cars.

1   Anderson et al. (2016). Autonomous vehicle technology: A guide for policymakers. Retrieved from https://www.rand.org/content/dam/rand/pubs/research_reports/RR400/RR443-2/RAND_RR443-2.pdf

ANY POLITICAL BENEFITS?

Self-driving vehicles are likely to feature in the future of human transport. The use of self-driving vehicles is also likely to have a few effects on government revenue, with revenues generated from speeding tickets, towing fees and DUI's eliminated from the publicly.[1] It would seem that AVs can be viewed from two perspectives. The first is that government revenues will be affected because, due to the probable implementation of AV technology, would lead to a waiver of those revenues due to an increase in the number of compliant drivers, many of whom would be classified as AV drivers. However, and in sharp contrast, the waiver of these potential revenues to any government could be a blessing in disguise due to some of the gains made by this type of technology. Because the technology has been successfully modelled as a safer system than the current human driver model, increases in safety of the system (whilst mitigating its inefficiencies) would save governments and taxpayers significant amounts of money.[2]

SAFER DRIVERS THAN HUMANS

Over a span of the past several years, Google's driverless vehicles have racked up more than 1.7 million miles, most of which were covered over a 6-year testing period. During the time, the cars have been involved in less than a dozen accidents, none of which were attributable to internal system failures.[3] Assuming that these findings on safety of AVs would be expected to replicate at scale, AVs would provide us with vehicles that are not only able to protect us during an accident, but also promises that the system will try and avoid the accident altogether. Whether the system can deliver a product of this calibre remains to be seen. However, there does seem to be enough in the technology to suggest that these technologies will be able to significantly reduce the number of accidents on the road, a fact that is likely to lead to other knock-on moments that may arise from this motor.

Note: it is entirely possible that the scaling up of a project of this nature may distort the trend line of such technology, thereby skewing our understanding of the technology. However, given the number of hours under analysis by Google in as far as AV is concerned, we are, at the very least, expected to continue advancing the technology as a species, and seeing what we eventually use once this advancement is achieved.

1   Desouza, K. C., & Fedorschak, K. (2015, July 7). Autonomous vehicles will have tremendous impacts on government revenue. Retrieved from https://www.brookings.edu/blog/techtank/2015/07/07/autonomous-vehicles-will-have-tremendous-impacts-on-government-revenue/

2   Gonzales, R. (2016, September 19). Government says self-driving vehicles will save money, time, lives. Retrieved from https://www.npr.org/sections/thetwo-way/2016/09/19/494648888/feds-to-set-rules-on-self-driving-vehicles

3   Desouza, ibid

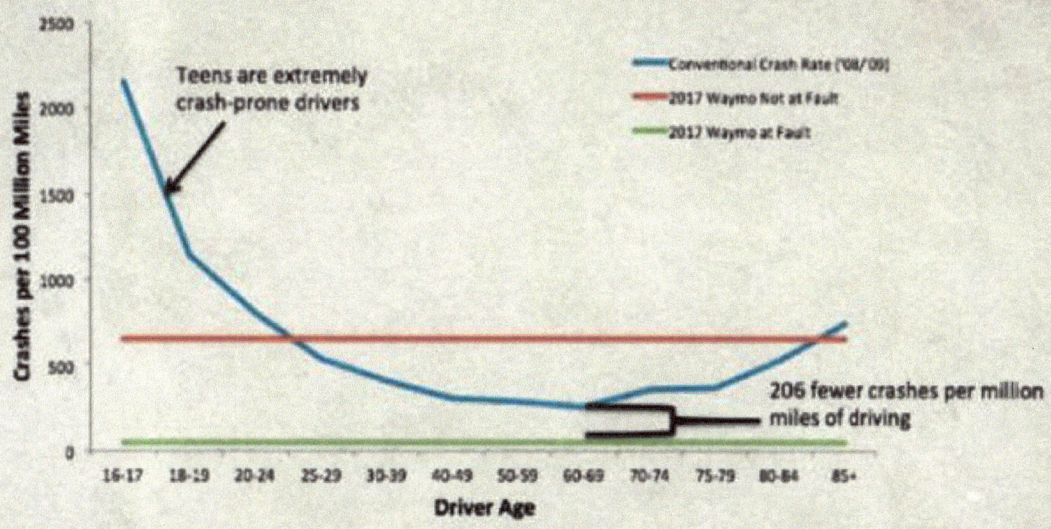

Conventional crash rate versus Waymo AV

According to figure 3, it should be noted that where the driver is at fault for the crash in either vehicle (conventional or autonomous), the autonomous vehicle improves safety even at the safest level of conventional crash rates.

LOWERED PUBLIC COST

Not only are AVs most probably going to make driving a safer experience for the human race, it should be noted that one of the key deliverables from an AV system is that it should be able to deliver safer driving statistics than human beings at driving, with the difference in degree a key determinant to how safe an AV is. The safer an AV system is, the safer it is likely to be. This safety is most probably going to reduce the number of fatalities from car accidents, point of which should be made. US public funds, as per a report by the NHTSA, were approximated at 7% of all motor vehicle crash costs, all of which cost tax payers around US $18 billion in 2010.[4]

If AVs achieve their primary goal (to fix the unreliable part of the driving experience, the human), then it can be reasonably expected that AV will lead to a drastic reduction in the number of accidents on the road, with these expected to trend towards zero. This would mean that governments would not bear the costs associated with said crashes. Thus, not only will AV make driving a safer experience all round, but it will also reduce expenditure by governments on accident related

4   Blincoe et al (2015, May) The economic and societal impact of motor vehicle crashes, 2010. Retrieved from https://crashstats.nhtsa.dot.gov/Api/Public/ViewPublication/812013

costs, and therefore reducing overall public costs.[5]

In addition to gains made from the overall reduction in the public spending on motor vehicles accidents, AVs are also expected to rectify many of the inefficiencies in present transportation systems around the world. In fact, the Brooking Institute estimates that present-day inefficiencies such as road damage, congestion and unrealised safety improvements are a waste of valuable resources. Because US transportation systems have lagged behind, a report by Winston and Mannering (2014) suggests that policymakers responsible for the improvement of public highway performance have failed (rather consistently) in their efforts to implement available highway technologies, and that the private sector would eventually implement beneficial highway technologies.[6] This report singles out technological innovations such as the driverless car as primed to leapfrog public sector and rewarding road users with the benefits of such technologies. Some of the benefits associated with driverless systems include:

- Expansion of roadway capacity and reduction in congestion through the use of GPS systems that can route vehicles through traffic more efficiently. Due to the interconnected nature of AV systems, congested can be cleared by diverting a certain percentage of vehicles off the highways and onto streets that may not be as busy, thus cutting travel times, reducing fuel wasted in traffic, as well as improving overall productivity[7]

- Diversion and adjustment of routing patterns for truck in order to avoid collision with critical and vulnerable infrastructure, thus cutting costs and preserving the lifespans of critical roadways and bridges

- Reduction/ elimination of transportation-related inefficiencies through the use of vehicle to vehicle communication technologies that would allow vehicles to navigate intersections of roads from all angles without having cars crashing into each other, thus effectively eliminating wait times at intersections.[8] This would work most efficiently in areas where most of the vehicles on the road would be AVs

SHIFTS IN TAX REVENUES

Traditional government revenues, as discussed in earlier portions of this text, would be reduced in as far as sources related to human driving errors. AVs, given their potential to reduce or eliminate altogether those human related driving errors, driverless technologies and innovations will lead to the elimination of such revenues. Further, trends that are associated with driverless technologies, such as

5   KPMG (2017, February). Impact of autonomous vehicles on public transport sector. Retrieved from https://assets.kpmg. com/content/dam/kpmg/ie/pdf/2017/07/ie-impact-av-vehicles-public-transport-2017.pdf

6   Winston, C., & Mannering, F (2014, June) Implementing technology to improve public highway performance: a leapfrog technology from the private sector is going to be necessary. Economics of transportation, 3(2), pp. 158-165

7   Desouza, ibid

8   KPMG, ibid

transportation economies (Uber and Lyft are leaders within this product category) are expected to shift the current car ownership business model to one with decreased vehicle ownership.[9] Decreased ownership, coupled with new trends within the automotive industry (pairing of autonomous technology with electric drive trains) are likely to further affect local government collections on traditional revenue streams, such as taxes that technological firms may be exempt to.

Overall, the report by RAND, as earlier referred to, suggests that policy makers should use safety (particularly if AVs are safer than average human drivers) as the guiding principle behind a decision on whether such technologies should be permitted on roads. In the case where AVs are granted said permissions on the basis of their safety when compared to the average human driver, policymakers would have to ensure that safety regulations and liability rules should be designed around this guiding principle, and then improved as the technology itself is improved. In short, policymakers should not only consider the best possible outcome for their constituents and community members, but also the state of technology and its safety relative to the average driver.[10] It goes on to suggest that it would be more prudent if policies around AV reflected its capabilities as currently constituted, and perhaps not where we all wish it could be, perhaps in a few years. Policy makers should consider that, as with any other technology, this one will probably require several iterations in order to improve the product and get it closer to full scale implementation based on improved functionality and safety standards. Expecting that the system will improve to attain such a level without necessarily being allowed into society in order to continue improvement may be ambitious.

AVs are a growing and evolving field in and of itself and would thus require dynamic policies that would allow it to thrive without necessarily putting any humans in any more harm than they would have if they had been driving themselves. Similarly, it is difficult to tell whether the best approach would include fundamental redesigning of AV systems in order for them to take the 'leap', or whether iterative improvements would be the best method to actualisation of the technology within society. However, the potential benefits of Level 4 and 5 autonomy, assuming that market failure is not achieved, outstrip some of the risks discussed here, and it will be interesting to see the direction that policy makers take within the context of ensuring that the technology is advanced whilst keeping the general populace safe.

9   KPMG, ibid

10   Anderson, ibid

Where we are going, we don't need any roads

Back to the Future II
1989

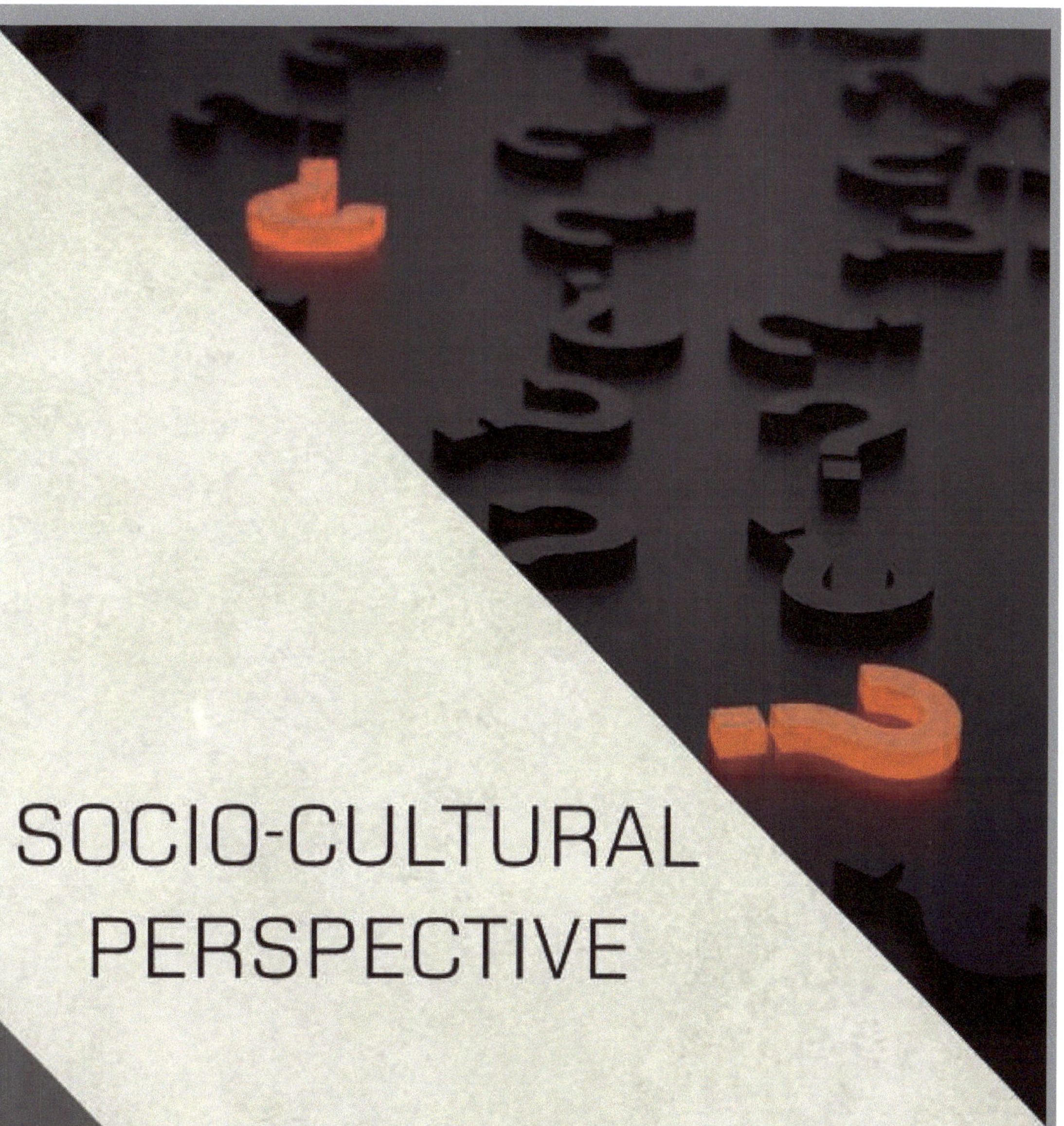

SOCIO-CULTURAL PERSPECTIVE

CHAPTER 08

SOCIO-CULTURAL PERSPECTIVE

Ultimately, uptake of AVs will firstly and primarily be driven by whether the general public considers them a safer alternative to human directed driving. As we have established in earlier sections of this book, humans are considered by most of those associated with the automotive industry as the key contributor to accidents globally. In fact, international evidence suggests that human error may be a factor in as many as 90% of global car accidents (crashes), adding that road user distraction or inattention is considered a contributory factor in around 10-30% of road accidents. Although these statistics do not lay blame on the driver as the cause of the accidents, it does suggest that drivers are a predominant factor in most accidents.

SAFETY

As has been mentioned several times already, the primary consideration for autonomous driving is that they will be safer than conventional human directed driving because they take away the most unreliable part of the driving process, the human. Indeed, it would be difficult to make a case for AVs if they do not reasonably perform better than humans, and it is yet to be established if this will be enough to satisfy doubts around the technology. Again, though studies continue to show that people are warming up to the idea that AVs could become mainstream (and that people will increasingly want to use them with time), it is difficult to establish the criteria with which consumers of these vehicles will use when making the purchasing decision.

Undoubtedly however, society can expect to benefit from improved safety on the roads, which will invariably lead to the preservation of life, as well as an improvement of its quality. A reduction in the number of accidents (including the intensity of the accidents) will be welcome and most likely improve the strength of the case put forward by AV technologies. However, because the industry is still relatively new and thus untested for the most part, it is unknown if consumers will measure the safety of AVs against human performance, or whether a much higher standard will be expected of the vehicle. As pointed out earlier, AVs may be expected to perform at a much higher level than human beings, a contentious issue. Humans will expect that the vehicle will protect them (while onboard) as well as place importance on ensuring that passengers and pedestrians are also protected from the technology.

Early indications from the AV world on the safety of their vehicles and the underlying technologies are mixed but mostly positive. Google's attempt at self-driving technologies has seen its AVs log more than a million miles with incidents few and far between; even so, the incidents have all been attributed to human error and not the technology. The same can be said of most other companies attempting to pioneer the technology, and this includes Uber. However, as much as the overall indications on the level of safety that the technology presents, it is not without its controversy (hint: Uber). New reports on the accident involving an Uber autonomous car that struck Elaine Herzberg in Tempe, AZ indicate that the car's software 'saw' the woman in its path yet 'decided' to do nothing, thereby eventually hitting and killing her. Early explanations for this 'decision' made by the vehicle software is that the vehicle was 'tuned' quite low as relates to the car's ability to detect hindrances on the road that posed no threat to the car, such as pieces of cardboard on the road. These types of incidences call into question the overall safety protocols written into driverless software, highlighting some of the risks associated with

AVs. Would humans rather a lower accident rate that is not up to them (in terms of their deciding their fate) or rather a higher accident rate in which the driver is considered in charge of their destiny because they are in control? It is difficult to say, and this may be one of the sticking points for the technology.

Such incidences are likely to take the shine off of the allure of AVs, but in the grand scheme of things, not likely to make a significant difference in their adoption, especially if it can be demonstrated that these technologies will present significant benefits. Further, these technologies are expected to contribute significantly to the overall reduction in road accidents, with simple levels of autonomy (such as electronic stability control: level 1) will likely reduce accidents occurrence by around 20%. Therefore, though some of the figures around expected safety of these systems may well be exaggerated or currently misunderstood, there is truth to the fact that there we are likely to witness significant strides in safety figures achieved in the automotive industry as a direct result of the adoption of AVS.

MOBILITY FOR AN AGING SOCIETY

The world is having to deal with a shift in demographic toward an aging society. Transport systems as currently instituted may not be robust/ adequate enough to deal with the unique and ever-increasing demands that an aging populace places on society. This may affect older road users, leading to complications around their mobility, its sustainability and most importantly, their safety. Some of the challenges that countries around the world have to consider when assessing the potential challenges (with regards to transport) associated with an aging population include:

- Better access and options of mobility solutions that cater to the elderly and allow them to meet their everyday transportation needs;

- Disability solutions that allow them to overcome their limitations as part of their daily lives;

- Ensuring that they are not isolated from society and can continue to engage in their work or social relationships; and

- Safety. Assessing the deterioration of driver capabilities with the onset of age, as well as their overall driver vulnerability.

In theory, AV technology is primed to be able to solve most if not all these challenges. In an effort to tackle the challenges brought about by its rapidly aging population, Japan has pioneered experiments in a rural town 71 kilometres north of its capital, Tokyo, on self-driving shuttle buses that are intended to help the aging Japanese population stay mobile and connected within their society.[1] Ideally, the adoption of these buses would be so that the elderly would be transported to and from banking, retail and medical services that, were it not for the buses, would have been unavailable to these

> *"Smaller towns in Japan are growing even faster than cities, and there are just not enough workers to operate buses and taxis...but there are a lot of service areas around the country, and they could serve as a hub for mobility services."*

Hiroshi Nakajim, DeNA, automotive software manufacturer

These buses will shuttle the elderly between a service area and a health care provider, moving at a speed of just 6 miles per hour, highlighting that safety tops its list of priorities. AVs promise to make societies far more accommodating to the lifestyles of the elderly and the disabled, especially since we are cognizant that we will all inevitably have to experience this for ourselves. This is especially true consider-

1   Chang, L. (2017, September 12). Japan focusing on a curious demographic with driverless cars - senior citizens. Retrieved from https://www.digitaltrends.com/cars/japan-self-driving-bus-elderly/

ing that a report published in the Journal of Gerontology found that senior citizens generally outlive their ability to drive safely by between 7 - 10 years, meaning that they risk social isolation once they stop driving. Add to this statistics that suggests that in the US alone, 560,000 people with disabilities never leave their homes due to the difficulties they face around transportation. In turn, this limits their opportunities for work and limits their productive output and contribution to the economy, a situation that isn't helped by their increasing dependence on caregivers.[2]

AV technology is not only expected to be a safe mode of travel for those that do not have the ability to drive, but it would also help the elderly in attempting to add to the number of years they stay within society's social circles. Though Uber and Lyft may well be considered viable alternatives for the elderly in society, such services are not readily available in many rural areas, and even so, it may be too expensive a service to be used as a daily option for travel. AVs are expected to lower the cost per mile of travel, meaning that transportation would be available to a wider audience.

"The aging of the population converging with autonomous vehicles might close the coming mobility gap for an aging society"

Joseph Coughlin, Massachusetts Institute for Technology Age Lab

The United States, for example, would appear to be a society that would benefit from the adoption of AV technologies for the elderly in their society, should the AVs deliver their promised potential of course. The population of people aged 70 and older is expected to increase to 53.7 million in 2030, not to mention that 16 million people aged 65 and over live in communities with poor or non-existent public transportation. These figures are expected to rise given that baby boomers mostly live in the suburbs and are 92% likely to not want to move. It is clear that a solution that goes to them is one way of thinking about ensuring their inclusion in transportation.[3]

2   Lahart, D (2017, January 6) Transforming transportation for the world's aging population and people with disabilities. Retrieved from https://www.ibm.com/blogs/age-and-ability/2017/01/06/transforming-transportation-for-the-worlds-aging-population-and-people-with-disabilities/

3   Chapman, M. M. (2017, March 23) Self-driving cars could be boon for aged, after initial hurdles. Retrieved from https://www.nytimes.com/2017/03/23/automobiles/wheels/self-driving-cars-elderly.html

"...carriages are pulled at the enormous speed of 15 miles per hour by 'engines' ...The Almighty certainly never intended that people should travel at such breakneck speed."

Martin Van Buren
Governor of New York
1830

CARS WILL CEASE TO BE PART OF ONE'S IDENTITY

In order to understand why this is so, we must first try and understand some of the motivations for buying a car, especially the importance a car held for previous generations. The car has, for countries of the developed world such as the US, been a part of the quintessential teenage experience. That is to say, the driver's license has always been a rite of passage for the American teen, as is with most other countries that permit driving between the age band of between 16 and 18 years old. In addition, the car, over and above the driver's license, held a very high position of value too teens, regardless of the type or condition of car. The car represents the ultimate expression of freedom and independence whilst also giving these teenagers an opportunity to sample a little bit of responsibility as an adult. However, with the introduction of AV technology, the need for personal ownership of a vehicle will diminish due to the simplicity of a ride-sharing model.

We should also consider the present mechanisms or systems in place that allow these teens to achieve these expressions of freedom and self. According to Fast Company, older teens and young adults no longer need cars in order to achieve this sense of self and freedom, meaning that the significance that the car plays within moulding the identity would reduce drastically. In short, there would be no need to attach significance to a vehicle because the need to own one would be diminished anyway. Therefore, vehicles may be ranked by the service offerings that a company offering ride sharing may offer in totality, that is, the quality of the vehicle as well as the bolt on services that make the package. The relevance of brands may also take a dip, as consumers would expect more than just the brand of the vehicle.[4]

4   Stewart, B. (2016, June 21) Autonomous vehicles: 5 predictions of social impact. Retrieved from https://www.linkedin.com/pulse/autonomous-vehicles-5-predictions-social-impact-brent-stewart/

COMMUTES WILL EITHER BE PRODUCTIVE OR RELAXATION TIME

Think of the autonomous driving experience like a train ride, minus the mayhem involved with actually getting on the train. For most of us, the train ride is mostly a pleasant experience, serene even. The choices afforded to a passenger on a train commute would be similar to a train ride; one may choose to either get started with the day ahead (or finished with the previous day's work unless the boss has your head) or enjoy a relaxed commute for some serenity before the hustle and bustle of the day begins. Either way, both options present much better options to manually driving the vehicle. The vehicle would therefore become an extension of either of these two things, or both. You could easily catch up with work emails and the business paper to get a fair understanding of what may be expected for the day ahead, and then sit back and relax, all whilst on your morning commute to work.[1] High speed Wi-Fi connectivity will probably come as part of the AV's standard features, will allow the car to connect us to colleagues, clients and customers (either way, these can still be achieved by people that use their phones to work as they will not be engaged in driving anyway) or actually provide us with interactive entertainment, information and news. Commutes are likely to change fundamentally.

EMPLOYMENT SHIFTS ARE TO BE EXPECTED

As discussed in earlier sections, adopting AVs is likely to mean significant reductions in the number of jobs available within the transportation sector. You will recall that truck drivers make up, as a single demographic, one of the largest employment groups in the US. It would be fair to assume that this would be replicated across the world, and this would mean that a substantial number of people would likely eventually be replaced by AVs. Further, even though it is likely that truck drivers would still be required (in addition to their roles as drivers, they also check that the vehicle is running well and make 'handyman' repairs to the trucks to ensure they get to their destinations), their roles would likely diminish in the long run. This does not even take into consideration the number of supporting industries around the transportation sector that would also be affected by the adoption of AVs. If anything, this technology will be disruptive.

However, it is important to note that the advancement of technology more often than does not replaces ultimately workers, but rather, actually increases the number of jobs available. To understand this, it is important to realise that, many times over, the threat of new technologies and the mass replacement of staff do not bear out with time. For example, the NASDAQ Composite Index peaking and subsequently crashing in the early 2000s was met with worry from the Association for Computing Machinery due to, at the same time of the crash, the global spread of the internet that enabled offshore outsourcing of software production. This led to worries that IT jobs would disappear en masse and what this would mean for the future of computer education and employment in the US. The last decade has borne out the worries as the US is a major player within the software development sector globally.

"Since the dawn of the industrial age, a recurrent fear has been that technological change will spawn mass unemployment. Neoclassical economists predicted that this would not happen, because people would find other jobs, albeit possibly after a long period of painful adjustment. By and large, that prediction has proven to be correct."

In essence, it is difficult to say what the net position on employment will be with the advent of driverless technologies. As previously highlighted, driverless technologies are likely to improve productivity, meaning that global economies would benefit economically. Further, it is difficult to say what the reaction of companies will be to the benefits of AV technology on their own improved productivity. What is almost certain is that AVs will require new industries to support them. Other initiatives such as retraining programs for legacy workers that have no other skills outside their previous work, more public-sector jobs, stronger unions and higher minimum wages. This will help those affected by culls to their jobs would most probably be able to transition to other jobs, particularly those that have already been earmarked as most likely to be replaced.

CONCLUSION

Society has tough decisions to make. What will happen to truck drivers is the most probable question, along with all the other jobs that are directly tied to the transport industry, for now. The full impact of the driverless economy is yet to be fully understood or appreciated, and it is thus difficult to surmise the net outcome of its adoption. Over and above this, it is estimated that the US economy alone will benefit from a $ 7 trillion by 2050 as a direct result of embracing autonomous technologies. The question that begs itself is therefore clear: do we protect the jobs, or take advantage of the significant benefits that the technology presents?

One approach to this existential question would be a shift in perspective: should we protect the jobs, or the people behind those jobs? Change is constant and unforgiving, just ask Kodak, among many other companies that were slow to react to changes within their industries, found out rather unceremoniously. It may therefore be an unfeasible approach to assume that humanity can reasonably expect to protect jobs that may be phased out eventually. It really is a matter of when at this point. It may therefore be unwise to think that protecting jobs would be a go-to strategy. An interesting approach may involve the constant training of people to fit into newly created jobs when they do come available. This would involve closely following market dynamics and shifts, identifying negative and positive trends within the employment market. The transfer of knowledge and skills does not only mean that societies will be able to aptly respond to the dissolution of jobs within their respective economies, but a coherent strategy around ensuring that the jobs are retained within the economy is critical.

Societal changes are not expected to be limited to job losses. Autonomous driving holds many prospects, especially when considering gains in efficiency. Consider an AV that picks your mother up for Thanksgiving dinner, and even better, drive her back to home. Consider also, the mother on a school run in the morning, amongst the million other things that moms are expected to execute on a daily basis. Perhaps the use of AVs, if considered safe at the time of their roll out, can be used to safely drop off and pick up the kids from school, in addition to dropping a fresh lunch off for them? Think of the societal changes that this type of technology may be able to introduce, both good and bad. All in all, society is likely to benefit from these technologies, because at the base of the idea behind it is a need to improve current modes of transport, especially considering that humans are the weakest link when considering the weaknesses of the current state of driving. We will have to wait and examine the developments that are almost a surety at this point. In the meantime, we can only sit and speculate on the future of AVs, with early indications pointing to significant benefits for society in store.

"Autonomous Vehicles" is more than a book, it is perhaps the most researched and up-to-date knowledge base on the topic of future transportation. This book does not only cover the hypothetial ideas and plans of technical minds; rather, it covers the topic in intricate detail. It includes topics like the technical prospects, the legal remifications of any autonomous accident, effect of autonomy on economics and what ethical concerns will arrise with the advent of autonomous vehicles?

The writer has expertly compiled noted research, indepth studies, and various technical research papers into one easy to read book!

As a bonus, chapters on environmental prospectives, political hurdles and socio-cultural effects of autonomous vehicles are also included to provide the reader with wholesome knowledge!

$19.95
ISBN 978-1-7320258-7-5
51995>
9 781732 025875

www.ingramcontent.com/pod-product-compliance
Lightning Source LLC
Chambersburg PA
CBHW042005080426
42733CB00003B/14